Environment and Economics in Nigeria

Routledge African Studies

1 **Facts, Fiction, and African Creative Imaginations**
Edited by Toyin Falola and Fallou Ngom

2 **The Darfur Conflict**
Geography or Institutions?
Osman Suliman

3 **Music, Performance and African Identities**
Edited by Toyin Falola and Tyler Fleming

4 **Environment and Economics in Nigeria**
Edited by Toyin Falola and Adam Paddock

Environment and Economics in Nigeria

Edited by Toyin Falola
and Adam Paddock

Routledge
Taylor & Francis Group
NEW YORK LONDON

First published 2011
by Routledge
711 Third Avenue, New York, NY 10017

Simultaneously published in the UK
by Routledge
2 Park Square, Milton Park, Abingdon, Oxon OX14 4RN

Routledge is an imprint of the Taylor & Francis Group, an informa business

Typeset in Sabon by IBT Global.

First issued in paperback in 2013

Library of Congress Cataloging-in-Publication Data

Environment and economics in Nigeria / edited by Toyin Falola and Adam Paddock.
p. cm. — (Routledge african studies ; 4)
Includes bibliographical references and index.
ISBN 978-0-415-89592-7 (hardback)
1. Sustainable development—Nigeria. 2. Nigeria—Economic policy. 3. Nigeria—Foreign economic relations. 4. Nigeria—Environmental conditions. I. Falola, Toyin. II. Paddock, Adam.
HC1055.Z9E5365 2011
333.709669—dc22
2011008737

ISBN13: 978-0-415-84742-1 (pbk)
ISBN13: 978-0-415-89592-7 (hbk)
ISBN13: 978-0-203-80749-1 (ebk)

To
Dr. Ademola Babalola of Obafemi Awolowo University, Nigeria
for his kindness and generosity

Contents

PART II
Oil and Politics

PART III
Environmental, Sociocultural and Political Dimensions

Tables

Figures

Abbreviations and Acronyms

AGIP	Azienda Generale Italiana Petroli
ATMN	Amalgamated Tin Mines of Nigeria
BDCP	Bioresources Development and Conservation Programme
BPU	Birom Progressive Union
BSCF	Billion Standard Cubic Feet
CAN	Christian Association of Nigeria
CFP	*Companie Francaise de Petrolieres*
CO2	Carbon Dioxide
CTM	Consolidated Tin Mines
DPR	Department of Petroleum Resources
ECOWAS	Economic Community of West African States
EIA	U.S. Energy Information Administration
EPL	Exclusive Prospecting License
FDR	Franklin Delano Roosevelt
FEPA	Federal Environmental Protection Agency
FME	Federal Ministry of Environment
GEAR	Grant to Enhance and Advance Research
IFAD	International Fund for Agricultural Development
IMF	International Monetary Fund
InterCEDD	International Center for Ethnomedicine and Drug Discovery
JCC	Joint Consultative Committee
JTF	Joint Task Force
LGA	Local Government Area
MEND	Movement for the Emancipation of Niger Delta
MDG	Millennium Development Goals
MLRU	Mine Lands Reclamation Unit
MNC	Multinational Corporation

MOSOP	Movement for the Survival of Ogoni People
MRA	Mine Reclamation Area
MUCDA	Mupun Cultural and Development Association
NAPEP	National Poverty Eradication Program
NARA	National Archives and Records Administration
NDES	Niger Delta Environmental Survey
NDDB	Niger Delta Development Board
NDDC	Niger Delta Development Commission
NDRBDA	Niger-Delta River Basin Development Authority
NDVF	Niger-Delta Volunteer Force
NGO	Nongovernmental Organization
NMC	Nigerian Mining Corporation
NNPC	Nigeria National Petroleum Corporation
NOSDRA	National Oil Spill Detection and Remediation Agency
Nsukka LGA	Nsukka Local Government Area
OEL	Oil Exploration License
OML	Oil Mining Lease
OMPADEC	Oil Mineral Producing Areas Development Commission
OPEC	Organization of Petroleum Exporting Countries
OPL	Oil Prospecting License
ORP	Oil Rivers Protectorate
RAF	Royal Air Force
RNC	Royal Niger Company
SAP	Structural Adjustment Program
SNC	Sovereign National Conference
SOCONY	Standard Oil Company of New York
SOHIO	Standard Oil of Ohio
SPDC	Shell Petroleum Development Company
UN	United Nations
UNDEDSS	United Delta Energy Development and Security Strategy
UNICEF	United Nations International Children's Fund
USAID	United States Agency for International Development
WHO	World Health Organization
WWI	World War I
WWII	World War II

Preface

Environment and Economics in Nigeria argues that issues relating to development and economics are closely connected to environmental resources; a largely extractive Nigerian economy means that perennial questions about environmental knowledge and who should benefit from an abundance of natural resources dominates historical and contemporary Nigerian society and politics. Environmental awareness represents an increasing international concern that affects the developing world in a variety of ways and Nigeria offers a divisive example of the contentions between environmental conservatism and desires for long term economic growth. The oil industry figures prominently in the chapters that follow owing to its post-colonial importance to Nigeria's revenue stream and the attending environmental strain oil companies place on local inhabitants and ecology. Yet, the chapters that follow are not limited to the affects of multinational oil companies, but they reflect a variety of environmental issues as well as how Nigerians, historically and contemporaneously, view their environment. Contestations over access to natural resources are not new but the players have changed overtime. In the pre-colonial era local rivalries or resistance to the slave trade reflected strategies used by local groups to evade or combat hostiles, strategies in which the environment figures prominently. The colonial era represents a period where British colonial environmental ideals clashed with indigenous environmental knowledge, often culminating in a benefit to the British export trade and occasionally resulting in devastating affects to local environments. Finally, the post-colonial period finds contentions arise between local people and the national government and local people versus the interests of multinational corporations. Each party stands to either gain or lose much.

Few works focus on environmental history in Nigeria. Nigeria boasts a wide range of books on economics and economic history such as Antony Hopkins's *An Economic History of West Africa* and for the post-colonial period, Toyin Falola's *Development Planning and Decolonization in Nigeria*. The purpose that follows is to shift the focus from primarily the economic issues to the environment and how access to resources or how natural resources have been used over time and who controls how the environment changes. The chapters that follow are based on research presented at a major international conference held at the University of Texas-Austin (March 27–29, 2009) on Science, Technology, and Environment in Africa. This volume brings together thirteen of those essays related to environmental issues and debates in Nigeria.

Broadly conceived as a debate between economic development and environmental conservation, the volume seeks to show how often, in terms of the developing world, the choice becomes one or the other. The chapters also reflect other contrasts. Several of the early essays begin by giving a broad overview of the economic and environmental issues; but by the end of the volume, the focus shifts from global and international influences to the individual perceptions that Nigerians hold regarding their environment and their place within it. Nigeria is also a picture of the extreme with oil affluence on one hand and starving local inhabitants trying to fish from polluted waters on the other. It is also about imperialism and trying to understand where its influence begins and where it ends. British imperialism changed the flow of trade, introduced new forms of transportation, and replaced food crops with cash crops. Did these changes begin with indirect rule or did the processes begin earlier with the transatlantic slave trade? Did independence and the national project end imperialism or are its vestiges still present in international economics and politics? How have all of these things shaped the Nigerian environment, which itself possesses contrasts of urban splendor and squalor, high-rise buildings and lush forests, dirt paths and major highways, all within walking distance of one another.

We want to thank all of the contributors who braved adverse weather conditions, long distances, and economic challenges to be with us in Texas in March 2009. Participants and presenters engaged in lively discussion over a three-day period, which without question the chapters that follow benefited from greatly. We are also grateful to Emily Brownell who coordinated the conference and a host of other graduate students (Tyler Fleming, Roy Doron, Kwame Essien, Sylvester Gundona, Saheed Aderinto, Tosin Abiodun, Segun Obasa, Charles Thomas, Jessica Achberger, Jason Morgan, and Nana Amponsah) and many staff at the University of Texas (Tony Araguz for his administrative assistance and numerous staff from several departments who chaired panels). The organizations and departments that supported us financially include the Department of History; The Warfield Center for African and African American Studies; The College of Liberal Arts at the University of Texas; The Frances Higginbotham Nalle Professorship; The Tacquard Lectureship; Alfred Fayemi, Hetty Ter Haar, Micheal Sharp and Ann Albuyeh; and the Alliance for Earth Sciences, Engineering and Development in Africa (AESEDA): A Penn State Institute.

Toyin Falola
Adam Paddock
April 10, 2010
Austin, Texas

Introduction

Environmental Themes in Nigerian History

Toyin Falola and Adam Paddock

In the last decade environmental history experienced a surge in popularity worldwide, especially in areas of study associated with the developing world and twentieth-century imperialism. Part of the interest in environmental history stems from growing awareness and concern about ecological problems, but perhaps more important is the buzz surrounding globalization. Environmental studies dispense with traditional nation state boundary concepts on which history writing often relies. In that vein, some of the most enterprising studies of environmental history analyze the global exchange of ecology along with their negative and positive effects on local environments. A large number of these studies focus on the role of imperialism as broker between transfers of ecology from one region to another. Yet economic issues cannot be excluded from a discussion of environmental history, particularly in terms of the developing world.

The following volume focuses on environmental issues in Nigeria and how they relate to economic development and political problems. In fact, environment and economics are two themes present throughout Nigerian historiography. Nigeria's economic dependence on extractive industries and agriculture production over the last several centuries underscores the important connection between economics and environment. The importance of these themes to Nigerian communities remained throughout the eighteenth to twentieth centuries but what changed was how Nigerians and British officials implemented different strategies for dealing with ecology and economy. Post-independence political and economic issues in part reflect a contestation over rights to resources, how those resources should be used and who should benefit.

The chapters that follow do not suggest solutions for all of the issues in Nigerian history related to economy and environment but they do represent a variety of problems and perspectives for the colonial and post-colonial periods. The introduction that follows will offer an overview of environment in Nigerian history beginning with the pre-colonial period. It will also address some of the issues between Nigerians and colonial officials over the use of the environment and how each group sought different goals.

Finally, we will address contemporary environmental issues in Nigeria and the political problems these issues tend to ignite.

The goal is to understand how economic and environmental change are interdependent in Nigerian history. Whether those changes are characterized as beneficial or harmful depends on the value judgments of the reader. Whereas Nigeria as a political entity did not exist until the early twentieth century, the term is used here to describe the geographical region that would later become the Nigerian state for simplicity in identifying the general area under discussion. The artificial nature of the Nigerian state also raises questions about how this might relate to contestations over access to environmental resources in contemporary Nigeria as different groups make various claims on environmental resources.

The historiography of environmental history and theoretical concepts concerning the role of the environment in history generally falls into two categories. Early theoretical concepts favored explanations that used environmental conditions as a determinant of physical and racial characteristics. These early ideas about the role of environment on people were used to support racial ideologies and to justify the supposed superiority of European cultures over Africa. Yet once the influence of the environment on human characteristics waned, the new historiography focused on how humans used techniques to control the environmental landscape to suit their needs or how human decisions lead to long-term ecological changes. Within this broad model, there are two main groups; one argues that "man lived in a constant state of war against nature until he was introduced to modern agriculture and industry through the imposition of western science and technology," whereas "others argued that pre-colonial Africans lived in harmony with their surroundings."[1] Jennings notes that these two views describe extremes; however, environmental historians currently accept a hybrid of the two positions.[2] What elucidates the discussion is recognizing that each group, Nigerians and British Imperialists, sought to influence the environment, but their strategies differed owing to varying economic and social goals.

Environmental history is also closely related to dependency theory and the concept that the structure of a central metropole and periphery explain the political and economic exchanges between colonizers and the colonized. Dependency theory explains the economic relationship between Britain and Nigeria to a point owing to the predominantly extractive characteristic of trade. Dependency theory falls apart in the sense that it focuses on the ability of the metropole to dictate the terms of trade. This concept has some merit but it fails to acknowledge the ways that Nigerians adapted to the environment and how they initiated changes in colonial policies. In many ways, an argument can be made that Nigerian economic strategies dictated responses from the colonial government.

Theories regarding the interaction between environment and humans or how characteristics passed from generation to generation informed much of the early work on environment and racist ideology. As early as 1809, Jean Baptiste Lamarck wrote that individual characteristics were passed onto children during their lifetime. Lamarck believed that adaptation to the environment depended on how these characteristics were transferred from one generation to another.[3] In many ways, colonial concepts that supported interpretations of the cultural inferiority of Africans depended on this ideology. The British civilizing mission suggested that with education they could change African cultures by infusing a new set of characteristics that would enable African societies to mirror Western views of what civilization ought to look like.

Social Darwinism dominated many ideas about race and inferiority in the mid to late 1800s. Social Darwinism represented a departure from Lamarkian ideas in the sense that characteristics were not passed on during a lifetime but were biologically hereditary. The middle to late 1800s and early 1900s advocated biological determinism where characteristics were determined by biology and blood, not environmental circumstance.[4] Yet, after the turn of the century, sociologists began to recognize the affect environment had on the success or failure of individuals. Grand ideals of the ability of the individual to rise above their circumstances were mitigated by the harsh realities of poverty and lived experiences. Friedrich Ratzel and Ellen Churchill Semple popularized the theory of environmental determinism, which argues that the physical environment is the main determinant of culture.[5] Either directly or indirectly, the basic concepts of environmental determinism informed many of the colonial era misunderstandings of West Africa.

Several writers of pre-colonial and colonial Africa suggest that one of the problems of western Africa related to the influence of a tropical climate on the characteristics of Africans. This body of work suggests that Africans failed to be more successful with development because the climate reinforced poor work values and other limitations. Some British colonial officials believed that the natural environment in West Africa led to laziness. They also believed that men were lazier than women partly owing to the types of work women pursued such as farming, carrying heavy loads of goods and selling in markets. Yet not all researchers interpreted the affects of environment on characteristics in the same ways.

Despite the work of numerous African historians, the ideology of environmental determinism appears to be resurfacing under a new guise. Jared Diamond's *Guns, Germs, and Steel* offers a more complicated view of modernization theory but at its core it arrives at some very similar conclusions.[6] It is still based on the basic assumption that European dominance in the eighteenth and nineteenth century depended on climatic and geographical conditions, which stunted African development. Therefore race, according to Diamond, was not the cause for European military might but a complex formula of environmental and social factors; however, Diamonds argument still suggests the inevitability of European dominance. In fact, in many ways Diamond is popularizing a colonial ideology about climate and development from the early 1900s.[7] Diamond, a biological evolutionist, fails to consider that perhaps African development sought to value different goals other than expansion. In fact, because Africa is commonly believed to be the origin of humanity, the continent is largely missing from the debate. Historians have shown that political maneuverings and treaties were just as important as military technology in European conquest.[8]

Gilberto Freyre in *The Masters and the Slaves* argues that not all Africans were the same but depending on the African environment in which they lived some Africans were better slaves than others. Freyre suggests a connection between environment and characteristics: "There was their greater fertility in hot regions. Their taste for the sun. Their energy, always fresh and new when in contact with the tropical jungle."[9] Therefore, in Freyre's view, Africans had a biological predisposition to be slaves in a tropical climate and part of the reason for this depended on their connection to tropical African climes.

> Slaves were brought in large numbers from other African cultural areas as well, and many of these areas were superior to that of the Bantus. Brazilian society in its formative stage benefited from the best that African Negro culture had to offer; it absorbed those elite elements, if one may put it that way, which were lacking in the same proportion in the Southern United States.[10]

Freyre's point was to suggest that the Brazilian populace was composed of a superior hybrid of culture compared to other groups such as African Americans. His goal was nationalist in scope, but in terms of environment, Freyre suggests a connection between the physical environment and human traits.

British colonial officials and researchers advocated their own set of stereotypes and misperceptions based on civilizing mission ideology and economic exploitation. Justifications for British Imperialism relied on the assumption that Nigerians did not possess the knowledge to adequately manage their environment for maximum economic output, and without British guidance Nigerians would remain a backward people. Historians of African history over the last sixty years have countered and disproved many of these mistaken ideas about Africans, the African environment and the complexities of the West African economy prior to European imperialism. We now turn to an examination of African historiography that offers a clearer analysis of the relationship between Africans and their environment.

PRE-COLONIAL NIGERIAN ENVIRONMENT AND ECONOMY

People outside of Africa held a variety of misperceptions about Africans in general and many of these misperceptions have their origin in a misunderstanding of pre-colonial Africa. The following section addresses several of these misunderstandings and how those ideas relate to environment and economy. Oddly enough, many misperceptions regarding Nigeria continued long after the colonial period and still resurface in the works of some researchers from developed countries. We will review some of the key themes regarding the pre-colonial Nigerian environment and economy. These themes include trade, agriculture, money, political organizations and society and environment.

A central misperception of pre-colonial Nigeria included the idea that environment and society went unchanged for centuries and that Nigerians failed to develop civilization or were able to overcome the constraints of subsistent agriculture.[11] Work by archaeologists have shown that up to 8,000 years ago groups in Nigeria were using boats, making them some of the oldest known boats in the world, which may have contributed to the movement of people and trade along the southern edge of the Sahara.[12] In many ways, owing to how Nigeria looked different from Western civilizations carried with it an inherent stigma of inferiority; therefore, because Nigerians did not develop their environment in the same way as Westerners it meant that Nigerians failed to understand how to best utilize their natural resources. For many Westerners the perceived economic backwardness of Nigeria was explained by a lack of knowledge and environmental ignorance.

Nigerian and West African historians have shown over the last sixty years the complexities of the pre-colonial economic and political systems in West Africa. Regarding Nigeria, as early as the late 1800s Samuel Johnson documented the political organization of the Yoruba of western Nigeria, which was a complex system of city states and a growing kingdom before the advent of British Imperial occupation.[13] The Hausa in the North also had a highly organized society based on a West African interpretation of Islam.[14] The British colonial government recognized the political organization of the Hausa and used it to govern much of Northern Nigeria via their philosophy of indirect rule. Southeastern Nigeria presented a more complex problem for early researchers and colonial officials to understand. Research has shown that the Igbo of southeastern Nigeria were politically organized in small local groups that depended on decision making by groups of elders and not dictatorships by chieftains. In addition to local governing councils, the Aro, a group that controlled trade across the region represented another layer of economic and political influence.[15] All of this research suggests governance by using a variety of political organizations and not groups of disorganized people.

Even during the pre-colonial period, subsistent agriculture and extractive resources were an important part of the Nigerian economy. Environmental knowledge was closely tied to economic success or failure. David Northrup argues that as early as the 1600s West Africans made economic decisions based on market value and that they were not duped by European traders. "Such facile images of African weakness, gullibility, and victimization go hand-in-hand with exaggerated images of European dominance and control."[16] Cowrie shells from South America filled an important function in the West African economy by providing a means of liquid income. Because cowrie shells were not part of the West African environment, their degree of rarity gave them value as a means of disposable income.

African economic historians have also shown that the Sahara desert and the sub-Saharan tropical jungle did not inhibit long distance trade. The ability of West Africans to trade with North Africa and East Africa suggests they were not limited by their environment but knew how to adapt and use the environment to suit economic needs. Gabriel Ogundeji Ogunremi shows how Nigerians engaged in trade prior to industrial transportation systems developed by British colonialism. He argues that Nigerians were adept at using pack animals, rivers and other environmental and geographical features to their advantage.[17] Ogunremi's purpose was to show that Nigeria had a variety of transportations systems in place which made sense to Nigerian markets before the arrival of British colonialism. In addition, his research illustrates the connection between environment and economics in pre-colonial Nigeria. Nigerians placed villages, markets and trade routes in locations that made the most sense to their needs. In this sense, Nigerians modified their environment based on strategic economic and social interests.

A.G. Hopkins' *An Economic History of West Africa* remains one of the key texts on the subject. Hopkins argued that West African economies reflected free market economics in many ways and their economies were neither backward nor dysfunctional but represented a complex exchange between long distant overland, local and overseas trade. Yet Hopkins clearly relays the importance of environment and how Africans modified their environment to achieve economic goals:

> Shifting cultivation and other forms of extensive agriculture were found where some or all of the following conditions prevailed: a low population density; a shortage of fertilizer; and an insufficient variety of crops. Where population was sparse and land abundant, farmers realized that it was important to maximize output per man rather than output per acre. European observers, obsessed by the notion (derived from their own experience) that output per acre was a priority of universal applicability, failed to grasp the principles underlying shifting cultivation.[18]

Hopkins suggests that African goals for economic output in terms of the environment were similar to Europeans, but Africans adopted different strategies based on their knowledge of the environment and the ratio of human capital to land.

West Africans' knowledge about the natural environment was extensive in pre-colonial Africa. What missionaries often labeled witches or medicine men were people who specialized in local herbs, information that was handed down from one generation to the next. Nigerian historiography could use more work on this area of research. Some work has been done on the diaspora in this regard including the connection between herbalists and religion, such as voodoo. James Sweet shows that there was a close connection between diaspora religion, resistance and Catholicism.[19] Furthermore, Sweet's work suggests that not only did slaves from West Africa possess environmental knowledge about local herbs but they were able to adapt those skills to a new environment in South America.

Akinwumi and Filabi in Chapter 3 provide an analysis of pre-colonial military strategies that used knowledge of local herbs in warfare. In addition to herbal knowledge to make biological weapons such as poison, the authors show how Nigerians used the environment in many aspects of warfare. Whereas Nigerians may not have used the same terminology as chemists in the West, they clearly understood how to manufacture biological poisons and also how to make medicines from local herbs. The authors also highlight the importance of this knowledge in respect to slave raiding in the nineteenth century. Donald Omagu in Chapter 5 shows how contemporary environmental degradation related to oil extraction in the Niger Delta has destroyed many of the herbs that Nigerian communities depended on for hunting and medicine.

As important as environmental questions are to African economic strategies in the pre-colonial era, the environment cannot be argued as the sole determiner of political and economic decisions, but it was important in terms of what economic decisions Africans made, how they chose to pursue long distant trade and the ways that imperial nations sought to control African resources. Imperialism's primary goal was political control for economic advantages, but a by-product of the imperial project included significant changes in how the environment was used and it challenged Nigerians' relationship with their environment.

IMPERIALISM AND ENVIRONMENT IN NIGERIA

The imperial project in many instances was closely tied to economic goals; however, economics alone do not explain the aims of the British colonial office. Yet, economics and environment played key roles in the process of establishing a British Colonial Empire in West Africa. Mungo Park explored West Africa along the Niger River to prove which direction the river flowed. Traders and businessmen hoped that this information would assist them in trade with West Africa. At the time, British traders possessed little knowledge of the interior. Initially, it was the Royal Niger Company ran by Sir George Goldie that established a British trading and power base in Nigeria.[20] Robinson and Gallagher argue that the transition from British companies to formal colonial rule in Africa depended on strategic decisions regarding the

British colony in India.[21] Their thesis, widely popular in the 1950s, has now been widely disproved; however, actual British profits from colonies in West Africa during the late 1800s was perhaps less important than what businesses expected they could make from natural resources in West Africa.

The transition from trading company control to formal colonial rule in Nigeria largely relied on the fact that the Royal Niger Company did not have the resources to effectively control trade throughout the region. Local long-distance traders, such as the Aro, were able to resist the Royal Niger Company's attempt to control local markets. When the British Government took responsibility of the region from Sir Goldie, they gained control from local leaders and then began a "pacification" campaign to extend government control to rural areas. In Eastern Nigeria, this meant removing the Aro, other trade groups and secret societies. Colonial trade depended on political control so that disturbances between varying ethnic groups did not interrupt British trade. In the past, disputes between different groups could cause trade routes to shut down, which did not favor British interests.[22]

The effect of British economic policies on Nigerian agricultural practices is widely documented. Imperialism depended on not only political control but on refashioning the environment to maximize their economic interests. Transportation, urbanization policies and agricultural products are a few sectors of the environment-economic nexus by which British imperialism influenced changes in the Nigerian environment. These changes offered benefits to some but also long-term issues for many Nigerians. Yet Nigerians were not passive actors but sought out ways to benefit from British economic strategies and also to resist changes imposed on the environment by British economic policy.

British colonial officials placed a high priority on transportation. Transportation was needed not just for the movement of goods from the hinterland to the coast but also for the movement of troops to ensure political control. The British colonial office had to achieve its goals with limited funds. World War I, the depression of the late 1920s and 1930s and World War II all contributed to strict constraints on the finances of the colonial office. To develop and maintain transportation routes including roads and railways, colonial officials often depended on forced labor either by directly conscripting workings or by going through local leaders to acquire the needed labor.[23] Ekundare has shown that transportation development by the colonial government was designed to serve colonial trading interests. The routes of roads and railways directed goods in directions that served British interests and not the trade routes through which Nigerians transported goods.[24] Colonial transportation redefined the Nigerian landscape and initiated new waves of migration from rural areas to centralized urbanities.

Towns and cities existed in Nigeria before the advent of British colonialism but British officials reconstructed those spaces and added new ones. The development of Lagos as an urban center was largely owing to the needs of British colonialism. In fact, between the perceived benefits of the city and new modes of transportation, large numbers of Nigerians migrated from the rural areas to cities. The extent of this movement was so great that British officials expressed concern over the quickly growing towns and their ability to maintain the infrastructure. In the 1940s, the British colonial office proposed a plan of propaganda to try and keep the younger generations on rural farms instead of migrating to towns.[25] Migration patterns were part of shifts in thc rclationships bctwccn pcoplc and thcir cnvironment. Younger Nigerian boys and girls sought out the advantages of the urban environment in large numbers but colonial policy intended to limit movement to the best of its ability.

British economic strategies reflected the demand of Nigerian goods on the international market. Groundnuts, coffee and palm oil were valuable goods and palm oil was important to the British industrial economy owing to its use as a common lubricant for machinery in the early 1900s.[26] Yet the importance of these items to local economies was tertiary. Nigerian diet depended on yams, cassava and other foodstuffs. Profits for British colonial companies depended on the mass production of non-food items, which adversely affected the availability and cost of food for many Nigerians. British demand for these non-food items resulted in a change in the Nigerian environment by replacing local foods for groundnuts, palm oil or cocoa, which was sold on international markets.

It is important to recognize that the transition from local foods to items with an international demand did not occur immediately. It was a process in which British companies and Nigerian farmers engaged in over time. In southeastern Nigeria, Igbo women benefited greatly in the early 1920s from the palm oil trade. They even adopted strategies to maximize their profits by adding water to palm kernels to increase the weight and therefore the price they received from British trading companies. The Women's War of 1929 partly occurred because women were upset about the drop in the value of palm oil and kernels.[27]

Kairn Klieman in Chapter 1 relates the international connections between the United States, Britain and Africa during World War II and how petroleum transport routes became an important impetus for U.S. finance in Africa. In some ways, the global war led to post-war global economic relationships that were unheard of prior to World War II. Klieman shows that oil companies played a role in African economics long before the discovery of oil in West Africa. She also notes some kerosene consumption patterns by Africans in which they purchased small quantities at a time and used kerosene containers for a variety of household uses. It is not Klieman's intent to deal with what happened to the packaging of goods in terms of trash and environment but there could be plenty of work to do on this area for Nigerian social history.

Charles Ukeje argues in Chapter 10 that an evaluation of resistance and domination during the colonial period explains the persistence of ethnic conflicts and post-colonial issues in the Niger Delta. He broadens the definition of acts of resistance to include more than armed rebellion to show how ethnic groups used their culture and identity to reject colonial intrusion. Ukeje also shows how the colonial government used ethnic differences as a way to keep peoples in the Niger Delta divided. These divisions and conflicts continue into the present via various issues; therefore, Ukeje seeks to connect contemporary political divisiveness to a culture of ethnic resistance during the colonial period. This is particularly interesting considering that the primary goal of colonial rule was to facilitate economic ventures, in many ways, the same topic of dissent in the contemporary Niger Delta region.

Disease is an important subject in terms of the colonial experience in West Africa and the relationship between people and the environment. Malaria and other diseases were part of the reason that European intervention into Africa occurred when it did. Prior to the invention of quinine, the death rate of Europeans who traveled in sub-Saharan Africa was extremely high. Yet the invention of quinine was not the sole reason for European expansion. Even after advances in medicine, Europeans still did not pursue settler colonials in tropical Africa; however, the concept of the civilizing mission and humanitarian concerns of missionaries led to profound changes in terms of disease and medicine.

British colonial officials brought new strategies to the treatment of disease in Nigeria during the colonial period. In fact, British officials expressed significant concern about the potential dangers of infectious disease including malaria, plague, leprosy and yellow fever. British administrators relied on tactics developed in Italy and that were used in Europe to prevent contagion. They reshaped the human geography by establishing leper colonies and cordons to control much feared outbreaks of plague. More work is needed to evaluate how Nigerian communities dealt with disease in the pre-colonial period as well as how colonial strategies either augmented or interfered with indigenous disease treatment strategies.

Jamaine Abidogun in Chapter 11 addresses how perceptions have changed in post-colonial Nigeria in terms of health and science. She argues that Western education and Christianity have in many ways either replaced or reduced the role of indigenous medical knowledge. Not only does Abidogun use interviews to illustrate these contemporary opinions, but changes in how Nigerians view religion and health have broader implications. From the perspective of environment, the relationship between the people and the local ecology has changed. In the same vein, because Nigerians have replaced indigenous health knowledge with Western medicine; this has a direct affect on the economics of health care in Nigeria.[28]

ENVIRONMENT AND ECONOMICS SINCE INDEPENDENCE

The predominant issue in post-independent Nigeria has been how to maintain political stability and attempt to manage economic development.

Whereas contested governance and fractured political parties tend to dominate post-colonial Nigeria, environment and access to natural resources or the benefits of those resources has been a key factor in Nigerian politics and economics. Pre-colonial structures that governed land access were replaced by bureaucratic systems that tended to focus on national goals and not local economic conditions. Independence led to new opportunities for the Nigerian nation but it has also been riddled by unfulfilled dreams. Environment has played a critical role in post-independent Nigeria owing to debates regarding access to natural resources have been important since the early 1960s.

One of the perennial issues regarding Nigeria's post-colonial economy is the continued reliance on extractive industries. In Chapter 2 Fwatshak discusses the nature of post-colonial mining in Northern Nigeria. Tin and garnet mining supplied communities in Northern Nigeria predominantly composed of rural farmers with supplemental income during the dry season. He argues that mining continued to be an important part of the Nigeria economy in the post-colonial period but not without long-term affects to the environment and in the case of garnet, local conflicts over land access and soil degradation.

Deforestation is perhaps the most internationally known African environmental issue over the last twenty years. Scientists and environmentalists displayed great concern over the effect deforestation in Africa has on the biodiversity of tropical rainforests. Biologists expressed concern that deforestation could result in the extinction of numerous species of plants and animals in addition to questions regarding carbon dioxide levels and greenhouse gases. Most researchers largely contribute the disappearance of the rain forest to external markets for timber. Between the high demand for timber in Western nations and the poverty level in Africa, there are few reasons for Africans to decline exporting timber in large quantities; however, a variety of issues in Nigeria besides external markets contribute to deforestation. Areola argues that deforestation in Nigeria is the result of the actions of millions of individuals who seek out more agricultural land to feed a growing population, among some other issues. The pressures of a developing economy, poverty and political issues that have harmed local economies all influence how local communities interact with their environment, and in this case have led to extensive deforestation beyond the demand for timber from external markets.[29]

Extractive industries such as fishing and mining were important aspects of the Nigerian economy throughout the colonial period, but by the time Nigeria gained independence, oil began to assert itself as Nigeria's most important export commodity. Oil often emerges at the center of economic and political debates about corruption, politics, economics and environment. Klieman suggests in Chapter 1 that global economic relationships during the colonial period may have created a political situation conducive to economic corruption. She details the role of the French government in maneuverings to gain as many resources and development projects as possible in West Africa via the lend lease program. Klieman suggests that what contemporary politicians view as a particularly African form of corruption was practiced by French colonial officers in Africa before independence and oil companies played a central role in those events.

The oil industry is one of the most divisive issues in contemporary Nigerian politics. The affects on the environment and an uneven distribution of basic infrastructural improvements have lead to sporadic outbursts of violence and oppression by the government over the last forty years. Some argue that the importance of oil revenue to the national economy outweighs the negative affects to local environments; however, others suggest that a high rate of money laundering and corruption by government officials prevent adequate development in oil rich areas.

In Chapter 5 Donald Omagu provides an exhaustive list of the environmental effects of oil production on Niger Delta communities. Omagu argues that a combination of indifference by the national government and irresponsibility by oil multinationals have created a situation where the costs of oil production on local communities rises to absurd levels. Local communities suffer from the destruction of local means of food production, which exacerbates conditions of poverty and unemployment. The issue strikes directly at ideas about the values of self-governance and the responsibilities of representative government. He suggests that violence in the Niger Delta is not the cause of dysfunctional radicals but the final recourse for youths dispossessed of their land.

Samuel Oloruntoba argues in Chapter 7 that the central problem to environmental abuse in the Niger Delta is related to problems of governance. Either the unwillingness or inability of the Nigerian Government to enforce existing laws protects illegal activities of oil companies in the country. He also suggests that processes of corruption within the national government that were established during military rule continue although Nigeria is currently a federal republic. Therefore, the form of government appears to be less important than a set of relationships between oil companies and politicians that continue unchallenged. Oloruntoba also suggests that some of the tactics used by the government to control unrest in the Niger Delta reflect tactics used by colonial governments during the colonial era to control the politics and economics of African communities.

Dele Ashiru expands on the specific problems of Nigerian governance in relation to the Niger Delta in Chapter 6. Ashiru suggests that the problem originated in the process of decolonization where policymakers decided to ignore committees that were already issuing warnings about possible problems in the Delta region. He sees the problems of Nigerian government so pervasive that the only solution is to reorganize the Nigerian State and to restore a sense of representative government that includes transparency and responsiveness to all minority groups.

It is important to note that many nations in Africa, as Ashiru notes in Nigeria's case, were artificial colonial creations that combined a variety of ethnic groups into a single political entity. Similar tactics were used in Eastern Europe at the end of WWI and WWII, but in the 1990s many of these artificial boundaries fell apart; in some cases peacefully and at other times the result of civil war. So why does Nigeria remain a single nation instead of dissolving into several autonomous states? This is no doubt a complex issue that cannot be fully answered here but a shared experience of colonial exploitation and environmental resources partly account for the resistance to fragment the nation. Despite the ethnic differences and interregional conflicts, Nigerians on some level do seem to share a sense of nationalism. In addition, environmental resources, such as oil, support the national government because too many regions of the country would suffer economically from division. This was in part an issue during the Biafra Civil War.

In Chapter 12, Alexander Kure argues that whereas these other interpretations of the crisis in the Niger Delta are important, the people in the Delta face fundamental concerns over security. Kure uses three dramas to make a sociopsychological argument regarding how people view issues in the Niger Delta. Environmental degradation not only destroys the agricultural and fishing industries for local communities but perhaps it also causes deep underlying psychological issues related to the destruction of the land, the home for many communities for hundreds of years. Kure believes to address problems in the Niger Delta policymakers need to recognize fundamental problems of security in terms of crime and environmental protection so that the land can support future generations.

The actions of oil multinationals and their affects on the environment did not go unchallenged in the Niger Delta region. In Chapter 8 Michael Sharp introduces the reader to Ken Saro-Wiwa and his efforts to stir up an environmental movement to protect the Niger Delta region's environment and people. Ken Saro-Wiwa was one of several Nigerian activists, such as Wole Soyinka, during the 1990s that challenged corruption and championed freedom despite personal danger. Whereas the story ends tragically for Ken Saro-Wiwa, the costs in human life reflect the extreme deprivation and the extent of the damage to the people living in the Niger Delta.

Post-colonial Nigerian agricultural business strategies also include a variety of tactics to expand the production of cash crops by using outgrower schemes. One such attempt included growing barley in the late 1980s and early 1990s. Businesses hoped to use modern irrigation and fertilization techniques to grow barley to produce alcoholic beverages; however, various issues including social protests and weather problems led to a general failure of the plan.[30] The lack of industrial development and the extractive economic relationship between Nigeria and the international market continues to lead to new strategies to maximize the land for exporting agricultural goods and other natural resources.

A predominant question for Africa in general is why so many post-colonial governments have failed to industrialize their countries and improve infrastructure. Perhaps even more importantly, why does poverty still sporadically affect large numbers of Africans when Africa has such a large amount of natural resources and land? These questions cannot be fully answered here but it is advantageous to consider some possible avenues of inquiry as way of concluding thoughts on post-colonial Nigeria. The high population density in Nigeria is part of the reason for strain on the environment but it does not fully explain poverty; high population density and overpopulation are two different issues.

Researchers and activists most often blame corruption and government mismanagement for poverty and underdevelopment; however, there are other issues that need to be considered as well. Godwin E. Okurume shows that Nigerian farmers in the post-independence period responded to market forces. In short, when the demand for exported agricultural products was high, farmers produced fewer domestic foods and more exports.[31] It is possible structural issues in the Nigerian economy explain part of the problems associated with underdevelopment and poverty. International demand for agricultural export goods grown in Nigeria could be limiting the amount of domestic food grown because the international markets prices are more advantageous for farmers that the lower value of domestic food products. Yet, this does not absolve the Nigerian government from apparent mismanagement of national resources and revenue, but it does illustrate how local agricultural production strategies could be exacerbating problems of poverty.

CONCLUSION

Nigeria's primary source of wealth is in its abundance of labor and natural resources. Common Western paradigms from the 1960s to 1980s included alternate interpretations of the relationship between economics and politics, namely, the contestation between capitalism and communism; therefore, a large number of books were dedicated to economic issues in the developing world. Environment appears in these arguments only as a tertiary interest, but the extractive nature of the Nigerian export economy makes environmental questions a key subject of pre-colonial, colonial, and post-colonial history.

As the labels of historical division suggest, Nigerian history is often categorized in relation to the colonial experience. Without question colonial intrusion was an important event in Nigeria's history, but it is somewhat Eurocentric to place such prominence on the influence of Europeans on the development of the continent. Environmental history offers a thematic analysis of Nigerian history that needs more exploration owing to an opportunity to evaluate the *long duree* relationship between the people, the environment, and the economy. This introduction has sought to offer a cursory overview of how environment and economics in Nigeria have been related throughout history and that despite numerous stereotypes Nigerians were not the mere victims of their environment but they possessed detailed knowledge and utilized strategies to make the best use of their resources to achieve specific economic goals.

An additional topic that needs evaluation is the connection between culture and environment in Nigerian history. How people viewed their environment changed over time from the seventeenth to the twentieth century. Cultural changes, of which colonialism was a part, meant that the relationship between people and the environment was not static but a dynamic and changing relationship. Changes in religious beliefs, science and economic practices have affected the relationship between people and their environment. Debates in Western nations suggest that this is not specific to Nigeria,[32] but how Nigerians deal with development is partly tied to their specific cultural ideals about the responsibilities of people in terms of their environment.

The chapters that follow offer a wide range of interpretations and case studies of environmental and economic issues in Nigeria. They by no means encompass a full range of possible topics, but they do provide an introduction for how environmental history and contemporary environmental issues inform us about the relationship between Nigerians and their environment. The introduction claims that whereas environmental history is a fairly new topic, the close connection between economics and environment owing to the extractive nature of the Nigerian economy means that environment has been an important aspect of Nigerian historiography, although it often is not the central focus.

A.G. Hopkins concluded *An Economic History of West Africa* by noting, “The history presented here has sought to direct attention away from the adventures and triumphs of great leaders, past and present, and towards the activities of the overwhelming majority of Africans, those who have never ranked among the elite.”[33] Environmental histories of Nigeria could serve the same purpose and they do not necessarily need to ignore political and economic issues that remain central to the everyday lives of those who live in Nigeria. Environmental history reveals the actions and difficulties of average Nigerians and shows how they sought to shape the environment to meet their needs. Environmental and economic history are overlapping fields that can reveal the subtle but important political and social changes so important to everyday life in Africa’s most populous nation.

NOTES

1. Christian Jennings, “Chapter Seven: African Environments and the Colonial Era,” in *Africa: Volume Three—Colonial Africa, 1885–1939*, ed. Toyin Falola (Durham: Carolina Academic Press, 2002), 124.

2. Ibid.
3. Jean-Baptiste Lamarck, *Zoological Philosophy: An Exposition With Regard to the History of Animals* (Chicago: Chicago University Press, 1984). Lamarck's concept about how human characteristics are passed on has been disproved by biologists and psychologists, but his ideas were not uncommon in the middle 1800s. Lamarck's ideas were part of the early ideology of hierarchy between peoples that suggested that non-Christian pagans were lesser peoples that the Christianized West. This later gave way to social Darwinism and the concept that inferiority was a biological characteristic.
4. There is a clear distinction between the ideology of racial inferiority before the 1940s and after the 1940s. Lord Lugard wrote, "They vary in their mental and physical characteristics according to the amount of negro blood in their veins"; *The Dual Mandate in British Tropical Africa,* 5th ed. (London: Archon Books, 1965), 67. Racial stereotypes began to change, especially after WWII, which depended on environment instead of the idea of blood.
5. Ellen Churchill Semple, *Influences of Geographic Environment: On the Basis of Ratzel's System of Anthropo-Geography* (New York: H. Holt and Co., 1911).
6. Jared Diamond, *Guns, Germs, and Steel: The Fates of Human Societies* (New York: Norton, 2005).
7. Lord Lugard attributed the difference between tropical and temperate climates as part of the reason why European countries had larger nation states and one dominant organized religion. See *The Dual Mandate,* Chapter 2.
8. Adu Boahen, *African Perspectives on Colonialism* (Baltimore: John Hopkins University Press, 1987), 33–34.
9. Gilberto Freyre, *The Masters and The Slaves: A Study in the Development of Brazilian Civilization* (New York: Knopf, 1956), 282. Freyre's main point here is to show the superiority of African Slaves to South American Indians in terms of their ability to work in a tropical environment.
10. Ibid., 299.
11. The work of archeologists provide a lot of information on pre-colonial Nigeria and offer a beginning for discussions about environment and how people in Nigeria have utilized natural resources and how those strategies have changed over time. See Raphael A. Alabi, "Environment and Subsistence of the Early Inhabitants of Coastal Southwestern Nigeria," *African Archaeological Review* 19, no. 4 (Dec 2002): 183–201.
12. Peter Breunig, Katharina Neumann, and Wim Van Neer, "New Research on the Holocene Settlement and Environment of the Chad Basin in Nigeria," *The African Archaeological Review* 13, no. 2 (Jun 1996): 111–145.
13. Samuel Johnson, *The History of the Yorubas: From the Earliest Times to The Beginning of the British Protectorate* (Lagos, Nigeria: CSS Ltd., 2001).
14. David Robinson, *Muslim Societies in African History* (Cambridge: Cambridge University Press, 2004).
15. A. E. Afigbo, *The Warrant Chiefs: Indirect Rule in Southeastern Nigeria, 1891–1929* (London: Longman, 1972); Nwando Achebe, *Farmers, Traders, Warriors, and Kings: Female Power and Authority in Northern Igboland, 1900–1960* (Portsmouth, NH: Heinemann, 2005); Don Ohadike, *The Ekumeku Movement: Western Igbo Resistance to the British Conquest of Nigeria, 1883–1914* (Athens: Ohio University Press, 1991); and Gloria Chuku, *Igbo Women and Economic Transformation in Southeastern Nigeria, 1900–1960* (New York: Routledge, 2005).
16. David Northrup, *Africa's Discovery of Europe, 1450–1850* (New York: Oxford University Press, 2002), 51–53.

17. Gabriel Ogundeji Ogunremi, *Counting the Camels: The Economics of Transportation in Pre-Industrial Nigeria*, (New York: NOK Publishers International, 1982). For a more comprehensive evaluation of the Trans-Saharan trade, see A.G. Hopkins, *An Economic History of West Africa* (New York: Longman, 1973), Chapter 3.
18. A. G. Hopkins, *An Economic History of West Africa,* 35.
19. James Sweet, *Recreating Africa: Culture, Kinship, and Religion in the African-Portuguese World, 1441–1770* (Chapel Hill: The University of North Carolina Press, 2003).
20. For an overview of the transition from the Royal Niger Company to colonial rule see Toyin Falola and Matthew Heaton, *A History of Nigeria* (Cambridge: Cambridge University Press, 2008).
21. Ronald Robinson, John Gallagher, and Alice Denny, *Africa and the Victorians: The Climax of Imperialism in the Dark Continent* (New York: St. Martin's Press, 1961); for an assessment of Robinson and Gallagher see, A. G. Hopkins, "Resistance, Collaboration, and the Transformation of the Terms of Trade," in *Historical Problems of Imperial Africa,* ed. Robert O. Collins (Princeton, N.J.: Markus Wiener Publishers, 1994), 88–93; Roger Louis, ed., *Imperialism: The Robinson and Gallagher Controversy* (New York: New Viewpoints, 1976).
22. Ohadike, *The Ekumeku Movement.* Ohadike argues that pacification in eastern Nigeria was largely owing to British desires to control trade routes and politics so that trade would not be disrupted; S. N. Nwabara, *Iboland : A Century of Contact with Britain, 1860–1960* (Atlantic Highlands, N.J.: Humanities Press, 1977), 97–98.
23. Nwabara, *Iboland.*
24. R. Olufemi Ekundare, *An Economic History of Nigeria 1860–1960* (London: Methuen and Co., 1973).
25. Ibadan prov 1/1 2600, "Social Welfare, General Questions, Report of Juvenile Welfare in the Colony," National Archives at Ibadan; Oshun div 1/1, 667, "Report on Social and Juvenile Welfare," Sept 23, 1946, National Archives at Ibadan.
26. Robert Pearce, "The Colonial Economy: Nigeria and the Second World War," in *Development Studies and Colonial Policy,* ed. Barbara Ingham and Colin Simmons (Hoboken, N.J.: Taylor and Francis, 1987), 267.
27. Susan Martin, *Palm Oil and Protest: An Economic History of the Ngwa Region, South-Eastern Nigeria, 1800–1980* (Cambridge; New York: Cambridge University Press, 1988).
28. For a discussion on strategies for preserving indigenous medical knowledge and the related issues, see O. A. Osunderu, "Sustainable Production of Traditional Medicines in Africa," in *Appropriate Technologies for Environmental Protection in the Developing World: Selected Papers from ERTEP 2007,* ed. Ernest K. Yanful (Dordrecht, Netherlands: Springer, 2008).
29. Olusegun Areola, "Deforestation in Nigeria: The Pressures at the Grassroots Level," in *Deforestation, Environment, and Sustainable Development: A Comparative Analysis,* ed. Dhirendra K. Vajoeyi (London: Praeger, 2001), 173–196.
30. G. Porter and K. Phillips-Howard, "Outgrower Barley Cultivation on the Jos Plateau, Nigeria: A Case Study of Agricultural Innovation," *The Geographical Journal* 160, no. 3 (Nov 1994): 319–327.
31. Godwin E. Okurume, *Foreign Trade and the Subsistence Sector in Nigeria: The Impact of Agricultural Exports on Domestic Food Supplies in a Peasant Economy* (New York: Praeger, 1973).

32. As early as Henry David Thoreau's *Walden,* and perhaps earlier, Western societies have questioned the consequences of development and how people are distanced from nature. Food, for example, becomes an abstract item where meat can be purchased, even without the bones, which distances the purchaser from the butchering process. Processed foods are just one example of how the relationship between environment and people change in response to economic developments. Social history may benefit from looking at how Nigerians over time have changed their perspective of environment and the extent to which colonialism influenced those changes.
33. A. G. Hopkins, *An Economic History of West Africa,* 296.

Part I

Environment and Development

1 From Kerosene to Avgas
International Oil Companies and Their Expansion in Sub-Saharan Africa, 1890s to 1945[1]

Kairn Klieman

INTRODUCTION

In August 1944, in an effort to inform its employees of the vital role it was playing in the Allied effort to win World War II, the Socony-Vacuum Oil Company (later to become Mobil) published an article in its newsletter entitled "The Saga of the Seven Engineers."[2] It detailed the "African adventure" of one of its long-standing employees, a Mr. Edward D. Seymour, who spent thirty-one years with the company, "most of it designing and constructing plants, terminals, and refineries in foreign fields." He was described as "an expert on pipelines, both land and underwater installations" and had rendered this service in both Africa and the Caribbean. Mr. Seymour's latest adventure began on August 20, 1940, when the United States Government called him and six other engineers—each nominated by the presidents of the leading U.S. oil companies[3]—to Washington, D.C., and asked them to draw up plans for an underground storage system that could hold a 100-day supply of aviation fuel and would reach from Takoradi, the deepwater seaport in the British Gold Coast, to Cairo. The job, which was expected to be completed in six weeks, evolved into a five-year engagement for Socony-Vacuum and included the building and operation of numerous oil terminals, blending facilities, truck delivery routes and supply facilities at West African coastal ports, airports and landing strips. Carried out in collaboration with Pan-American Airways, the United States Ferrying Command (later to become the Air Transport Command), and the U.S. Army Engineering Corps, this effort was initiated to provide supplies to the Allied forces in their battles with the Axis powers, especially in North Africa and the Middle East.

The establishment of this supply route was a result of months-long collaboration and planning between Winston Churchill and Franklin D. Roosevelt.[4] With the fall of France and the resultant threat to Allied supply routes in the Mediterranean, the need for a fully functioning route across the continent of Africa had become an imperative. However, given the harsh climatic conditions through which the route was supposed to run ("up tropical rivers, over deserts, and through jungles" as one article described), the creation of this African supply route was no small feat. It was also, in its earliest stages, carried out in full secrecy, not only to prevent strategic information from being leaked, but because it was in direct violation of both U.S. and international law. By working to establish what came to be called "The Takoradi Route" Roosevelt chose to override the isolationist mandate of the American public, the Neutrality Acts passed in 1939 and international laws that would have allowed Hitler to declare war on the United States for carrying out an overtly hostile act.[5] The gamble paid off in the end, for it was through the route that supplies were channeled to Britain's Eighth Army, allowing it to defeat Erwin Rommel's Afrikakorps in North Africa in November of 1942, thereby turning the tide of the war and ensuring the Allies continued access to Middle Eastern oil.

Although the strategic role of the Takoradi Route is mentioned in numerous histories of WWII, it is generally treated only briefly and with little detail as to its operation or effects in Africa. Only two scholarly studies exist, these being Culbert and Dawson's 1998 *Pan Africa: Across the Sahara in 1941 with Pan Am* and Deborah Wing Ray's 1975 article entitled "The Takoradi Route: Roosevelt's Pre-War Venture Beyond the Western Hemisphere."[6] These works, along with a soon-to-be released film entitled "FDR's Secret Air Force" by John Schwally, focus largely on the role Pan Am played in Allied effort (Culbert and Dawson) and the unprecedented nature of the collaboration between Pan Am and FDR (Ray and Schwally). Ray also emphasizes the importance of the route after the Allied confrontation with Rommel in North Africa. After the bombing of Pearl Harbor, for example, the route served as "the most reliable, and almost only, way of getting planes and supplies to both China and the Pacific."[7] As she notes

> The first heavy bombers sent to General Douglas MacArthur in the Philippines took off from Sacramento, California, but flew by way of the Takoradi Route. During 1942 the trans-Africa route handled the heaviest transport and ferry traffic of all routes in the world . . . ferrying hundreds of pursuit planes and bombers to the desert Air force, Russia, and India. . . . It is therefore no exaggeration to say that . . . the trans-African route probably saved Egypt and markedly changed the course of the war in Russia, China, and the Pacific.[8]

Given the vital importance of this route to the Allied victory in World War II, one may wonder why it has not received more attention from Africanist historians. This may simply be owing to a lack of conventional African sources on the topic, because the majority of data so far has been culled from sources not usually associated with the field (i.e., Pan American Airways archives, interviews with its former employees and State Department Records not pertaining directly to Africa). Also rather unusual is the fact that none of these accounts discuss the role of international oil companies in establishing the route, a contribution that was fundamental to its success because Pan-American had never done any prior business in Africa. Given that the supply route was up and running within ten weeks, there is clearly a more local history to be told. This article, derived from a larger project focused on the history of international oil companies in Africa, is only a beginning. It is, like previous works on the topic, rather company-centric, largely due to the sources from which data has been drawn.[9] A fuller version, rooted in African people's experiences related to the creation of the Takoradi Route, remains to be told.

Despite these issues, a number of important insights regarding Africa's economic and diplomatic history, especially as pertains to relations with the United States, are elucidated by retelling this story through an "oil history" lens. For example, the events documented here suggest that it was essentially an oil company (Socony-Vacuum) that led the way for the United States' earliest entry into the African political and economic sphere It also documents, for the first time, the degree to which international oil companies penetrated African markets before the discovery of oil; the role these companies, especially Socony-Vacuum played in the creation of the Takoradi Route; and the ways that the U.S. government utilized Lend-Lease aid to help promote these companies interests in Africa as well as to accomplish much broader political goals. Data for this latter section derive from a series of State Department documents that illustrate the problems the U.S. government had in dealing with the "Free French" in Equatorial Africa, especially in terms of improvements it paid for at the seaport of Pointe-Noire.

By focusing on these three aspects of African oil history I seek to achieve a number of broader goals as well. First and foremost, I seek to place Africa within the standard narratives of energy/oil history and twentieth-century processes of globalization, both of which tend to treat Africa in rather cursory ways. Second, I seek to illustrate the benefits of reconstructing oil history in Africa. As will be shown, approaching the past through such a lens can provide important new perspectives on Africa's history, especially in terms of its relations with the West. Finally, by focusing on the history of oil companies and African governments during the colonial period (and before the discovery of oil), I seek to identify patterns and trends, events and decisions that help to explain the origins of the underdevelopment and corruption that so negatively affects oil producing nations today. It is my contention that "downstream" (i.e., marketing and distribution) histories are as important as "upstream" (exploration and production) in explaining the origins of these problems in Africa, and this article constitutes a preliminary effort toward explaining why.

VACUUM OIL COMPANY, ROYAL DUTCH/SHELL AND THE EXPANSION OF AFRICAN MARKETS BEFORE WORLD WAR II

Standard Oil and its African Affiliates

Key to the rapid establishment and great success of the Takoradi Route were the vast marketing, storage and transportation networks run by the two leading oil companies in Africa before World War II. These two companies were Socony-Vacuum (later to become Mobil) and Royal Dutch/Shell. Socony-Vacuum was formed through a merger of Standard Oil Company of New York (or S.O.C.O.N.Y. as per its telegraphic name) and Vacuum Oil, Ltd. in 1931. Vacuum had been a subsidiary of the Standard Oil Trust before the latter's dissolution under antitrust laws in 1911; acting on its own thereafter, but still benefitting from cooperation with the other Standard companies formed as a result of the break-up,[10] Vacuum became the leading marketer of petroleum products in Africa. The history of how this success was achieved is presented in this section. It is important as a new contribution to African history in its own right but is also essential for illustrating the level of experience and expertise FDR was tapping into when he called the "Seven Engineers" to Washington, D.C., in 1940. On the other hand, whereas the following account deals with Royal Dutch/Shell, its early (marketing) history in Africa is rather hard to discern; company archives are not open to the public and official company histories tend to focus on the regions where production took place (i.e., Southeast Asia, the United States, etc). Although it appears the company made inroads into Eastern and Southern African markets by the late 1920s, details surrounding the 1928 combinations (to be detailed below) suggest that Africa did not become a key focus of marketing efforts until that date.

As numerous histories recount, the Standard Oil Company was founded by John D. Rockefeller in the late nineteenth century. He began his career in the 1860s as a refiner of kerosene in Cleveland, Ohio, and made his mark by aggressively borrowing funds, buying up rivals and when need be, forcing them out by undercutting them on the price of transport. His goal was to dominate the industry, and he understood from early on that this would best be achieved not through the production of crude oil but through monopolistic control of refining and distribution. To this end he established a system of secret "rebates" with the operators of the Atlantic and Great Western railways systems, which carried crude from Titusville, Pennsylvania—the center of crude production at that time—to Cleveland. Through this arrangement Rockefeller received substantial discounts for shipping large quantities of freight and was able to force out or cut secret mutually profitable deals with nearly all of his competitors. His success was phenomenal. In 1870, when Standard Oil Company of Ohio was first formed as a stock-joint company, Rockefeller held 10 percent of the U.S. oil industry. By 1879 he controlled 90 percent of U.S. refining capacity, and by the mid-1880s his company had gained control of roughly 80 percent of the marketing side of U.S. business as well.[11]

At this time the company operated in thirteen different states and several foreign countries; recognizing that a lack in centralized managerial authority was harming operations, Rockefeller and his associates formed "The Standard Oil Trust" in 1882. Its headquarters were established at 26 Broadway in New York, and at its incorporation, The Trust was comprised of fourteen fully owned and twenty-six partly owned companies. Key among these was Standard Oil of New York, which from early on was charged with the task of expanding export markets abroad. Being based in New York it developed "a more sophisticated and international approach" than the other Standard companies,[12] and until 1899, it remained the more dominant of Standard companies in terms of its capitalization, banking capacities and supplying of senior executives for the central organization.[13] At the time the Trust was formed, more than 50 percent of the kerosene refined in the United States was consumed abroad,[14] and petroleum products in general ranked fourth in value of all U.S. exports, outranked only by cotton, breadstuffs and provisions. Kerosene accounted for more than four-fifths of the value of these exports in the early 1880s; however, virtually all of it was sold to export merchants or representatives of foreign importers along the eastern seaboard. Standard Oil sought to alter this pattern, focusing first on sales to British and European markets through the creation of the Anglo-American Oil Company (1888), and then on the large markets of the Far East (primarily China and Japan).

The latter expansion efforts necessarily involved Africa, because the routes of transport to the Far East were through the Suez in Egypt or around the Cape of Good Hope in South Africa. It is thus no coincidence that the earliest Standard Oil affiliate companies were established in these two regions. The most important of these in terms of global expansion was the Vacuum Oil Company. Originally based in a refinery business in Rochester, New York, Vacuum specialized in the sale of lubricants and was involved in secret dealings with Standard Oil before the formation of the Trust. Incorporated as a partially owned company in 1882, it was originally valued at $25,000—making it the least capitalized of all Standard's companies.[15] Nonetheless, it established offices in Montreal and Liverpool by 1885, Toronto, Milan and Bombay by 1892 and after 1894, built manufacturing plants and affiliate corporations in Europe and Russia as a means of avoiding the high tariffs imposed on imported lubricants in foreign lands. Vacuum's success was rooted in a unique set of marketing strategies, which, according to Hidy and Hidy, "set the pace for all other standard oil marketers in improving and applying sales techniques."[16] As the authors explain

> Vacuum oil was . . . second to no other Standard oil unit in the use of packaging, especially of attractively decorated cans of appropriate size

> for easy handling. Travelling representatives sold cylinder and spindle oils, which had greatly improved by 1899, to industrial establishments by sample. After the initial orders "mechanical experts" followed with advice as to the proper type of lubricants and with supervision of their application. Vacuum men could get no nearer the customer.[17]

These marketing techniques were employed by Vacuum and a number of other Standard companies as they expanded into overseas markets. Vacuum's success in these endeavors was made apparent in 1901, when its capitalization jumped from $25,000 to $2,500,000 on Jersey Standard's books.[18] For its part, Jersey Standard created The Colonial Oil Company the same year, in an attempt to expand its own markets abroad; this venture was not especially successful, and most of its affiliates were incorporated into preexisting Vacuum over the next ten years. Through all of these efforts, The Standard Oil Trust saw its overseas markets expand immensely; as William Herbert Libby, Standard's foreign representative declared in 1886, their efforts had allowed petroleum to "force its way into more nooks and corners of the civilized and uncivilized countries than any other product in business history emanating from a single source."[19] As is illustrated below, Africa played an important role in this early history as well, although the continent tends to be overlooked in standard oil histories until the era of production (i.e., 1950s).

Vacuum Oil in South Africa

Although company documentation suggests that Vacuum products were being marketed in Egypt from 1896,[20] Vacuum's first affiliate office in Africa was established in Cape Town, South Africa, in 1897. Founded by Misters Sandham and White, the affiliate was described as a branch of SOCONY but supervised from Vacuum's London office.[21] Sources describing these early efforts tend to be rather nostalgic in tone, for they appear as articles in company journals that were written for overseas employees many decades later. Nonetheless, they do provide insight as to the special problems early Vacuum marketers encountered as they moved into foreign lands.

For example, as a foreign company Vacuum was required to pay an annual tax of one shilling per £100 of registered capital to operate in Cape Province. Given that the company was capitalized at £25,000, Mr. Sandham and Mr. White presented the Revenue Office a check worth £2.50s. The Cape Revenue Commissioner, "convinced that oil business meant big business," refused to accept such a paltry amount until he received an affidavit endorsed by the British Consul confirming that the declaration of capital was correct.[22] A lack of knowledge about local climatic conditions hampered early business as well. Because there were no proper storage facilities, the first shipment of 365 wooden barrels of petroleum products (kerosene and lubricants) sat on the docks uncovered and unattended for weeks. However, the two agents were not aware that wooden barrels could not withstand the African heat. As a result the contents of the barrels were drastically reduced in volume, and the agents suffered a real loss. Iron barrels were eventually procured from the Cape Railways Company, and storage was arranged with the Imperial Cold Storage Company. The company's first orders were for one barrel of 600-W cylinder oil from "Small and Morgan's Sawmills, Dock Road" and ten 10-gallon drums of the same product from the "Sheba Gold Mining Company of Barberton."[23]

In these earliest days of the oil trade with South Africa supplies were shipped only on sailing craft, as sailors and port authorities were fearful of explosions when transporting petroleum products by steam. Vessels that had been previously used for grain shipments were employed, and they arrived at Table Bay, South Africa one-hundred days after leaving the U.S. refinery at Bayonne, New Jersey.[24] Once products began arriving from the United States in cases (rather than barrels), sales expanded and Vacuum offices were established in Durban, Port Elizabeth, East London, Kimberly and Bloemfontein. Vacuum salesmen set out from these locations to market their product, relying on "cape carts" for transportation rural locations and bicycles in the more urban areas.[25] In 1901 a certain American named Mr. C.C. Campbell established a branch of Jersey Standard's Colonial Oil Company in Cape Town as well; it operated independently until 1908, when it was amalgamated with the Vacuum branch under the name of Vacuum Oil Company of South Africa, Ltd. This name remained in use until 1962.[26]

As the new century proceeded the demand for kerosene declined and fuel oils came to dominate overseas markets Thus, whereas Vacuum Oil Company of South Africa continued to sell lubricants, it also specialized in providing fuel for internal combustion and steam engines, catering to customers operating steamships, diesel motors and factories. A full-page advertisement from 1913 boasted that Vacuum's Pegasus Oil was "used by South African Railways, New Cape Central Railways, the Defense Force and all Government Institutions."[27] It appears that the Vacuum company had a monopoly over the South African petrol market in these early years; historical accounts of Royal Dutch/Shell, which was eventually to play an important role in South African history, indicate that the company's efforts to start a marketing venture in the region were halted by the South African or "Anglo-Boer" War, which took place between 1899 and 1902.[28]

Vacuum Oil in Egypt and Northeast Africa

Vacuum undertook a simultaneous marketing campaign in Egypt. There, however, it was likely in competition with Royal Dutch/Shell, which was busy establishing shipping routes through the Suez to the Far East during this period of time (1890s).[29] In 1902 a certain Mr. C. Xippas was sent from Vacuum's Paris affiliate (Vacuum Oil Company SAF, which oversaw operations in France, Belgium, Holland, Portugal, Spain, Switzerland, Algiers and Egypt),[30] to lease land from the Ports Lighthouse Administration for a storage depot at Gabbery. In 1907 the same individual, acting on behalf of the main Vacuum Office in Rochester, New York, purchased land and built a second depot at Ghamra, and Vacuum went on to buy or lease property for storage facilities and offices—over the next fifteen years—at Assiut (1907), Samanoud and Benha (1908), Port Said (1909), Suez (1914), Dekerness (1915), Aswan (1916) and Alexandria (1922).[31] The company operated under the name Vacuum Oil Company until the 1931 merger with Socony. In 1952 it was reorganized as a separate subsidiary ("Socony-Vacuum Oil Company of Egypt") and in 1955 its official name was changed to "Mobil Oil Egypt, Inc."

In the earliest decade of operation many of Vacuum's products were transported down the Nile River in lighters (flat-bottomed barges designed for shallow waters) and then sold to local traders who carried them inland on the backs of camels.[32] In this way regular networks were established that carried American petroleum products into the far interior markets of the Sahara from as early as 1915. In the more urban areas expatriates were often hired to market the products. A 1909 correspondence between a New York office manager to a newly hired salesman for Vacuum describes the level of commitment required:

> The writer is expecting careful, intelligent, systematic and painstaking work from you, and if he is not disappointed you will give such attention to our business. In that event your ultimate success may be looked upon as sure. To represent us on the road in a way that will be satisfactory to yourself and to this company, however, is not an easy matter. It will call for all the tact, energy, and persistence you are capable of bringing to bear up on the subject. . . . Make it a point to call at the office [e]very day from 4:30 to 5:00 pm. Let us have full reports on every company whose business you are soliciting, making certain that you enter upon your report the correct name of every customer called upon. Bear in mind our material will be furnished on approval to any customer, but avoid as far as you can sending out smaller packages than the regular full and half barrel quantities.[33]

As Hidy and Hidy have noted, it was this type of corporate culture, focused on the employment of highly trained salesmen who strived to "meet the special needs of each industrialist," and provided free demonstrations along with expert advice, that allowed for the phenomenal growth of the Vacuum Oil Company worldwide.[34] Such success was apparent in Egypt, for by 1927 the Cairo affiliate became a key distribution center in Vacuum's global operations and employed seven hundred individuals, 60 percent of whom were described as "natives."[35] Removed from the directorship of the Paris office, it now served as the head office for twenty separate affiliates located across Egypt, Syria, Palestine, Cyprus, Greece and Turkey.[36]

Also under its direction was the maintenance of relations with hundreds of "agents," i.e., private companies that sold Vacuum products through their own general stores in northern and northeastern Africa. Such contracts were drawn up in Sudan in 1925 (the Contomichalos company), Ethiopia in 1927 (The Arabian Trading Company), Djibouti in 1927 (The Arabian Trading Company) and Eritrea in 1936 (C. Scapellate).[37]

Undoubtedly owing to differences in economic and infrastructural development, but probably influenced by racial prejudices as well, Vacuum marketing strategies put a great deal more effort into reaching the customer in Egypt than it did in the southern and western regions of the continent. A prime example can be seen in the company's correspondence school of the mid-1920s, which consisted of ten printed lessons in Arabic, "each copiously illustrated" and designed to instruct locals on "the fundamentals of motor car driving and operation . . . including comprehensive knowledge of the function of every part of an automobile and full information concerning the care of each part." There was no charge for the course, and as of its first season (1927), fifteen hundred "students" had taken part. Not surprisingly, these participants were among the elite, as made evident in excerpts from letters sent to thank the Vacuum Oil Company for this service. A certain Dr. Iskander Gorgui, for example, reported that "since taking the course and applying the instruction he has been able to notice a great improvement in the operation of his motor car"; a Mr. Alfonso Tawkik promised "to keep the instruction papers 'very preciously in my library'"; and Mr. Habib Zaki, a senator, wrote that he had "tried to find inaccuracies in the course" but had not succeeded.[38]

Vacuum Oil in West Africa

Vacuum was the first to company to market petroleum products in West Africa and remained the sole supplier until after WWI. In 1908 it began its official marketing efforts by appointing specific trading firms in each region to serve as consignment agents for its products (French firms in French territories, British firms in British territories, and so on). Prior to this Vacuum products had made their way into west African markets through the work of shipping firms established in Africa; these bought kerosene in cases in New York and then sold it at various locations along the West African coast. In the earliest days these traders bartered kerosene along with cotton cloth, sugar, flour and hardware in exchange for various African products (generally agricultural goods, minerals and gathered products such as rubber, copal, skins, etc.). By the 1920s kerosene was being widely sold in rural regions through the activities of small-scale African traders. Vacuum documents report that these were primarily women, who generally bought and sold one case of kerosene per day (a "case" consisted of two five-gallon tins packaged in a wooden crate) and that the product was transferred to bottles and sold in small quantities to individual households. The traders also made profits by selling the wooden crates and tin cans that packaged kerosene. As was the case in the Far East, Africans recycled these materials to make a great variety of useful household items (wooden furniture, kerosene lamps, drinking vessels, etc).

Vacuum's marketing activities in West Africa were originally directed from the Lisbon office, due to its proximity to the region and the fact that shipping lines to and from Portugal were already well established.[39] From 1907 kerosene was marketed under the local brand name "Sunflower Oil"; as time passed the market grew to include lubricants for marine, agricultural and industrial machines as well as motor fuel. The latter was marketed under the Egyptian brand name "Sphinx Oil"; signs advertising this product could still be seen in the streets of Accra in the 1960s.[40] Although imported petroleum products tended to be handled, stored and distributed by the contracting agents, Vacuum maintained a staff of some twenty individuals in West Africa during the 1920s, these being "expatriate sales personnel and local clerical help as needed for the checking and preparation of sales statistics."[41] It was not until 1928 that a Vacuum office was opened in West Africa; records do not indicate the exact location, but it was likely in Accra or Lagos because it was in these two cities that Vacuum first built storage depots through partnership with local agents.

The number of Vacuum employees in West Africa increased to thirty by 1938. These were now referred to as "technical engineers and inspectors" and served as liaison officers between the Lisbon headquarters and the "thousands of distribution outlets in West Africa."[42] Although they were assigned houses, offices, cars and servants, employment in West Africa was deemed a particularly arduous post because these officers were required to spend half their time on tour. This indirect system of marketing—whereby Vacuum placed very few employees in the field and all the storage, handling and sale of products was carried out by local agents—was retained until after WWII. According to company documents this was the most economic approach in a region where consumption of petroleum products was "exceedingly small, thinly spread out, and increasing at a slow rate."[43] As any visitor to West Africa knows the practice of selling small quantities of kerosene in bottles for household cooking and illumination has not died out to this day, even in countries that produce large quantities of oil for global consumption.

Despite the slow rate of market growth in West Africa, the Vacuum Oil Company took great pride in the degree to which its products penetrated the continent. In West Africa it was reported that—thanks to the efforts of the local trading firms that had "pushed inland in search of new markets,"—"the Company's business followed into the interior of Africa, until . . . European and native traders in every village or crossroads in West Africa" sold Vacuum products. By the 1930s the company boasted that its markets reached deep into the heart of Africa, for its products were being regularly sold at Maidiguru in northeastern region of Nigeria, 800 miles inland from the Atlantic coast. Vacuum employees also seem to have been obsessed with documenting the success of the company's distribution efforts. In an article from *The Sales Generator* of January 1923 it was noted that marketers were in the habit of sending photographs of Vacuum products found in obscure locations to the New York headquarters. Indeed, "so keen has been the interest in the query 'where next' that whenever a new headline appears in the newspapers, the "VOCO" enthusiasts inquire to find out whether the event was correctly lubricated, so to speak."[44] Accordingly, the author noted that "Gargoyle D.T.E. Oil Heavy Medium" was being used in the engine room that provided lighting for excavations in the tomb of Ramses X in the recently discovered Valley of the Kings at Luxor.[45]

Corporate Combinations and the Entry of Royal Dutch/Shell, 1928–1939

Whereas the demand for petroleum products in West Africa may have been slight before World War I, the situation was altered during the war and the subsequent global economic boom of the 1920s. African economies underwent dramatic changes during these times, largely due to the skyrocketing demand for vegetable and seed oils (palm oil, palm kernels, copra, cocoa and peanut).[46] These products were used not only to remedy the global shortage of fats, but also in the production of high explosives for the war.[47] Both African producers and colonial governments benefitted from this boom in trade. The increased wages accruing to Africans further stimulated local markets by heightening the need for imported and manufactured goods (including, presumably, kerosene), as did the arrival of a great many new European settlers seeking to profit from the economic boom. Meanwhile, colonial governments utilized increased revenues and taxes to build new roads and transportation systems. It was during this period that truck and car transport came to challenge railroads as the primary means of transport for export goods. Although railroads continued to carry and deliver goods in bulk, the new roads and trucks were able to reach into regions that previously had no means of getting agricultural goods to market. As a result, new markets were tapped, trade expanded and local peoples were increasingly integrated into the global system of trade. These trends continued well into the 1930s. In 1938 Socony-Vacuum noted that gasoline sales in West Africa had increased tenfold from those of 1922, and that trucks accounted for 90 percent of such sales.[48] Lubricating oils and heavy fuel oils were also in high demand because tractors and other types of large vehicles were being used in the newly burgeoning colonial industries focusing on the extraction of timber, gold, copper, tin, diamond and rubber.[49]

The 1920s also saw an increase in demand for petroleum products owing to the development of air transport routes to and within Africa. The British, working through Imperial Airways and Royal Dutch/Shell, pioneered these efforts, with Shell carrying out extensive surveys to produce the aviation maps that allowed for the earliest flights to take place. In the 1920s much of this work focused on the development of a Cairo to Cape route and a Khartoum to Lagos route.[50] The latter was the precursor to the Takoradi supply route, and according to Shell histories, was the most arduous to accomplish. Because early airplanes could only fly short-range and during daylight hours, a series of refueling stations and lodging facilities had to be laid out across the desert and rain forest regions south of the Sahel. Thus, from Khartoum airplanes went to El Fasher (in the Sudan), Maidiguru and then Kano (in northern Nigeria), and finally, to Lagos or Takoradi on the West African coast. As Ray has noted, the route remained largely undeveloped until the advent of World War II, for it lacked "radios beacons, communication facilities, adequate runways, repair shops, planes, pilots, and mechanics." In all the flight took seven days to accomplish, an estimated 20 percent of the planes that attempted it had crashed in the desert or the rain forest and only one hundred planes had successfully traversed the route by 1940. Of those that did, many had to be stripped down on arrival because their engines were full of sand.[51]

Although regular passenger-carrying flights were not well established in Africa until the mid-1930s, the advent of mail carriers in Africa was a great boon to the oil business as well. Oil companies competed to serve as suppliers of aviation gasoline ("avgas") and lubricants as each new air transport route was initiated. Vacuum, for example, boasted in a 1913 advertisement that its products had fueled the monoplane carrying South Africa's first airmail in 1911. Royal Dutch/Shell, however, was the leader in supplying this fledgling industry; through the establishment of Shell Aviation Services, a company that specialized in the production of aviation maps, the servicing of airplanes, the creation of airports and landing strips and the supplying of petroleum products, it came to hold a monopoly over the sale of aviation fuel in colonies like Nigeria by World War II.

This data regarding the post-war economic boom in Africa sheds a new light on the importance of African markets in a key event of global oil history—the establishment of the 1928 "Consolidated" and "As Is" or "Achnacarry" Agreements (to be outlined below). These constituted the first attempt by major oil companies to create a cartel that would control production, distribution and marketing of petroleum products worldwide. Whereas Africa is mentioned in historical accounts of these agreements, it is generally treated rather cursorily or primarily as an eastern extension of the Indian market.[52] The latter, which was third largest in the world at the time (after the United States and Europe), had long been dominated by a smaller British Oil Company called Burmah Oil.[53] As Yergin has noted, however, it was the Anglo-Persian Oil Company (later to become British Petroleum), suffering from a surplus of crude and lack of marketing outfits, that initiated the push toward consolidation. In the face of a global oil glut and increasingly brutal price wars, John Cadman, the director of Anglo-Persian, saw both India and Africa as regions where markets could still be expanded. The possibilities of Africa's newly burgeoning economy was likely to have been especially evident to Anglo-Persian because its major shareholder was the British government, a powerful colonial force in Africa and India.

The Anglo-Persian Oil Company was formed in 1908 after the discovery of vast oil fields in Iran. Whereas it had become a major player as a supplier of crude, it was severely hampered by its lack of access to global markets. As the main supplier to the Royal Navy during World War I, it had gained an edge in the Indian Ocean periphery regions that included East Africa. In attempting to move beyond the fuel oil market, Cadman was faced with the choice confronting his competitors head on—a tactic that would be exceedingly expensive and not guaranteed of success—or convincing them to enter into joint ventures whereby the markets could be "rationalized" and divided up. Cadman chose the latter and began by entering into a deal with Royal Dutch/Shell and Burmah Oil that pooled markets and facilities in India in return for serving as a main supplier of crude to the two companies and agreeing to refrain from any form of marketing in their terrain.

Cadman's next target, as Yergin has documented, was Africa, for which he proposed a 50/50 split of markets with Royal Dutch/Shell. The conclusion of this "African Alliance" was, however, a bit more complicated because American oil interests had long dominated on the continent. Yergin notes that resistance to this plan came from a number of corners of the British government, which as the major shareholder had to approve the deal. Key among these was the Foreign Office and the Treasury, both of which were fearful of alienating the U.S. Government by "instigating what would effectively be considered a 'war' on the interests of Standard Oil."[54] It was for this reason that Winston Churchill was also initially opposed to the idea of a combination. He eventually conceded to the plan, however, because his alternative was to ask Parliament for funds to initiate Anglo-Persian's marketing campaign in Africa. Churchill justified this move by stating that the combination would "in the long run be more in the interest of the consumer than cutthroat competition."[55] The first and most lasting result of this attempt at combination was the Consolidated Agreement of 1928, which Bamberg describes as essentially an "imperial carve-up" of Eastern Hemisphere markets between Anglo-Persian and Royal Dutch/Shell.[56] The region consisted of " . . . an expanse, roughly triangular in shape, between Cyprus, South Africa, and Ceylon , taking in a great swathe of British imperial territory in East, Central, and Southern Africa." As Bamberg explains

> The area was named after the Consolidated Petroleum Company, which was owned by the Company [i.e., Anglo-Persian Oil] and Shell in equal shares, and which was the exclusive vehicle for the ownership and management of all their marketing operations in the Consolidated area. Shell and the Company were entitled and bound to supply in equal

shares the crude oil and products sold in the Consolidated area, and the profits of Consolidated were split equally between the two parents."[57]

Although the West African region was not technically part of the Consolidated Area, the joining of forces between Anglo-Persian and Royal Dutch Shell allowed the latter to gain better access to markets in Britain's African colonies (Gambia, Sierra Leone, Ghana and Nigeria).[58] By 1936–37, the two companies began to explore for oil in Nigeria, with Shell as the operator.[59] This work was interrupted by WWII, and oil was not discovered until 1953.

Yergin has noted that in agreeing to Cadman's 1928 African plan, the British did not simply approve an economic carve up of their imperial territory.[60] More important was the fact that it encouraged a move toward combination and international collusion, an act that would have been unthinkable on the part of the U.S. government given the history of Standard Oil and the antitrust sentiments that prevailed. In essence the British government gave Anglo-Persian permission to approach Standard Oil, hoping that it might "seek out market arrangements with the Americans in order to 'allay their jealousies and show that we are not out to quarrel.'"[61] Thus, it was with his government's encouragement, and with the goal of finalizing the details of the African plan as well future international combinations, that Henri Deterding, chairman of Royal Dutch/Shell, initiated the next phase of the 1928 combination trend. In August of that year he invited the leaders of the major oil companies to a two-week "shooting and fishing" retreat at Achnacarry Castle in Scotland. Among these were Walter Teagle of Standard Oil of New Jersey, John Cadman of Anglo-Persian, William Mellon of Gulf Oil, Colonel Robert Steward of Standard Oil of Indiana and Heinrich Riedmann, Jersey's "chief man in Germany."[62]

The result of the meeting was a seventeen-page document referred to as the "As Is" or "Achnacarry" agreement that laid out plans for managing oil markets well beyond the African sphere.[63] Its ultimate goal was monopoly control along the lines that Standard Oil Company enjoyed at the turn of the century. Although the companies adjusted agreements and created new formulas in attempts to achieve this goal throughout the 1930s, it was impossible to achieve, for as Yergin explains, "all over the world, there were too many producers and too much production outside the 'As-Is' framework." Whereas the long-term impact of the Consolidated and Achnacarry Agreements in Africa has yet to be documented, it is probably safe to say that the former resulted in a marketing push that challenged the supremacy of Socony, Vacuum Oil and other American-based "fringe players" (Gulf Oil, Union Oil, Texaco) that were establishing themselves in Africa at this time. However, faced with desperate straits in the midst of the depression, the Achnacarry group began to secretly attack each others' markets, and "their efforts at collaboration were paralleled by pervasive distrust, wariness, and deep-seated rivalry."

As Yergin notes, "All during the 30s, as Jersey discussed cooperation with Shell, it was continually giving serious consideration to merging its foreign business with Socony so it could better take on Shell."[64] By 1938 Standard Oil of New Jersey formally withdrew from the failing combination, and with the outbreak of World War II entirely new kinds of collaboration, organized by and for the Allied governments in their efforts to win the war, came to rule the day. It is to that history, including the creation of the Takoradi Route, that we will now turn.

WORLD WAR II, INTERNATIONAL OIL COMPANIES AND THE TAKORADI SUPPLY ROUTE

As early as the spring of 1937 the British government was well aware of the threat that the German-Italian Alliance posed to their shipping lanes and supply routes in the Mediterranean. Out of fear that their access to the Suez Canal might eventually be limited, the Committee for Imperial Defense began to build up oil storage facilities in North Africa and along the West African Coast, these being supplied primarily by Royal Dutch/Shell.[65] This work was done on the assumption that traffic would be diverted around the Cape of Good Hope in South Africa and that British cruisers could be based in and refuel from West African ports. To this end a British garrison at Freetown, Sierra Leone, began to be reinforced,[66]and a new 200,000 ton storage facility for petrol, oil fuel and lubricants was built in Capetown.[67] The British also recognized the strategic role that the Old Imperial Airway's Takoradi-Cairo flight route could play, and thus, through the work of both the RAF and Shell, began taking steps to improve the route.[68]

With the fall of France to the Germans (May 10 to June 22, 1940) West Africa's strategic role in the war was greatly enhanced. Roosevelt and Churchill had been in close conversation regarding the possibility of a trans-African route as early as April of 1940;[69] by that point it was clear that Britain did not have sufficient resources to execute a war on the continent. In terms of oil supplies, "Royal Dutch had managed to set up a few small depots. . . . They were close enough together to serve bombers, but not fighters, and the difficulties of supply held even the bombers to a trickle."[70] Thus, in July 1940 Roosevelt gave the secret directive to Pan American Airways to begin modernizing airports in the Caribbean and Latin America.[71] It was from these regions that oil and other war supplies would be shipped, not only to aid Britain, but also "to put the United States into a position defensively sound and offensively flexible."[72] Roosevelt was determined that the south Atlantic air routes not fall into Hitler's hands, especially that from Natal (Brazil) to West Africa, which provided the shortest point of passage between the two continents.[73] The U.S. army had long shipped supplies to the Middle East and Egypt through these South Atlantic routes, but because of a lack of aviation refueling stations in the interior of the African continent, it was required to send its planes around the Cape of Good Hope and up the East African coast. In developing the Takoradi Route, as Ray has noted, "8000 miles of boat travel and possibly a month's time could be saved."[74]

The next month (August 1940) the "seven engineers" were called to Washington and asked to come up with the plans for the 100-day fuel storage system along the Takoradi Route. It is likely that Socony-Vaccum was assigned the lead role because during the same month it moved the headquarters of its West Africa division from Lisbon, Portugal, to 26 Broadway in New York, and also established seven new sub-offices in West Africa.[75] The initial contracts with Socony-Vacuum were initiated by Pan American, which itself was being paid through "a confidential contract with the war department, using moneys from the President's emergency fund."[76] It is thus likely that Socony-Vacuum was paid through these same funds.

Pan-American gave Socony-Vacuum two specific mandates. The first was to re-survey the airports and refueling stations previously put in place by Shell and the Royal Air Force (RAF). In the earliest stages of this effort an employee named Keller from the aviation division in Socony's New Orleans branch was employed. According to the *Fortune* article of February 1943 he began his journey in Accra and ended it in Iran, dispatching reports from each of the company offices he visited along the way. As a result, "while he was still on the way to Suez equipment was being assembled, and before he returned to the U.S the first shipment was on its way."[77] In this way the supply line was functioning efficiently within a few months. Socony-Vacuum officials continued to carry out these types of surveys as the war continued, the most famous taking place after the bombing of Pearl Harbor. To ensure that the route could handle the increased traffic, the aforementioned Mr. Seymour, his colleague by the name of Petty, and the vice-president of Pan American Airways crossed the continent four times in thirty-five days, covering a distance of more than 39,000 miles.

Pan-American's second mandate to Socony-Vacuum was to ensure the delivery of supplies to the field. Through this effort Socony employees (both African and expatriate) became involved in such diverse activities as clearing and maintaining roads, procuring trucks and spare parts from the United States, recruiting and managing laborers, building lodging facilities at airports and generally instructing newly arrived expatriate laborers how to "outfit themselves for service in regions where many necessities of American life were not to be purchased." Company documents specifically highlight the contributions of African employees in ground transport, especially in terms of truck transport. Truck drivers had to possess considerable skills because roads and bridges were rarely in top condition, fuel supplies and replacement parts were few and far between (and were thus carried on board) and repairs had to be dealt with en route. A 1942 article in the Socony-Vacuum newsletter praised the efforts of these men, noting that "The drivers, mostly natives of the region, know their roads thoroughly and tackle the impossible with sublime confidence in their ability and the ability of their sturdy American trucks to get them through."[78] Socony also paid adolescent boys four dollars a month plus meals to accompany the truck drivers on their journeys. They served not only as assistants to the drivers, but also carried out the difficult task of extracting trucks from the mud during the rainy season, when in many locations roads essentially disappeared.[79]

Although research is still being carried out regarding the contributions of international oil companies to the war effort in Africa, it is clear that their role was seminal. As the author of the 1943 *Fortune* article noted, oil had become material, and "international oil companies such as Socony were integral, functioning, virtual military branches in services of supply."[80] After Roosevelt's signing of the Lend-Lease Act (March 1941) and his official announcement that aid would be supplied to Britain through Africa (August 1941), the importation, storage and distribution of all petroleum products was controlled through the British Petroleum Board (est. 1938) and the Petroleum Administration for War in the United States (est. June 1941). The two governments quickly became the largest consumers of petroleum products in the world, and competition between oil companies was temporarily reduced. Along with the masses of African peoples, other commercial enterprises and the various colonial governments, oil companies were enrolled in a massive mobilization/militarization effort directed by Britain and the United States. Each of these groups was called on not only to supply the war and accommodate newly arrived soldiers and civilians, but also to build, refurbish and/or maintain military bases, seaport storage and shipping facilities, hospitals, railways, roads and military camps. These types of activities took place throughout West Africa, although the British focused on protecting and maintaining their key seaports at Freetown (Sierra Leone), Takoradi (Gold Coast) and Lagos (Nigeria) and airports at Bathhurst (Gambia), Lagos, Takoradi and Accra (Gold Coast). The United States, through the authority of the U.S. Ferrying command, focused their efforts on Takoradi and Accra (hubs of the Takoradi Route) as well as airports and landing fields in Lagos, Maidiguru and Kano (Nigeria) and Fort Lamy (Chad). Its main seaport was at Pointe-Noire in French Equatorial Africa (modern day Republic of Congo). The role of oil companies in improving this port will be discussed below.

It is important to note that the massive mobilization effort generated a certain degree of economic development in Africa during WWII. This did not generally benefit the majority of the African populace because as producers of agricultural commodities they were paid lower prices by wartime purchasing boards. However, as Asheley notes, by benefitting from the labor of tens of thousands of Africans (both employed and conscripted) the colonies themselves underwent "something of an industrial transformation."[81] Essential to these changes were new and improved infrastructures for the transport and storage of petroleum products. Whereas I have not been able to document to what degree (or even if) international oil companies profited from their involvement in the war effort in Africa, it is clear that the transformations engendered aided them in expanding markets and increasing profits after the war.

For example, the demand for massive quantities of petroleum products initiated a shift to bulk handling and storage, which lowered costs for oil companies by reducing the need for barrels, tins and the manpower required to handle them. Furthermore, the construction of the new bulk storage facilities and the infrastructure required to supply them was largely paid for through Lend-Lease funds, and in most territories (except those of the French; see below), such facilities were handed back to the oil companies at the end of the war. After May of 1943 the wartime distribution system also tended to eliminate the need for local agents, i.e., the commercial firms that had long served as "middlemen" in the distribution systems of West Africa.[82] This provided an opportunity for the oil companies to reconfigure their business practices and focus on selling their products to the customer direct. Because such efforts were heavily focused on key supply centers; however, it is unlikely that markets for petroleum products were expanded far beyond the essential supply routes during the war. Socony-Vacuum documents suggest such was the case, for in the ten years after the war the company's main business strategy was the extension of bulk handling and storage systems into the far interior regions of the continent, as well as establishing Western-style service stations in urban areas.[83] Finally, the mobilization effort and reduced levels of competition created a climate in which new types of collaboration could be tried out. As the details regarding the construction of oil facilities at Pointe-Noire will illustrate below, such collaboration was increasingly important to oil discoveries.

U.S. POLITICS IN POINTE-NOIRE: TOWARD A POLICY FOR AID AND COMMERCE IN AFRICA?

One of the distinguishing, and eventually rather problematic, peculiarities of the Lend-Lease program was that to cut red tape and mobilize rapidly, the U.S. government generally began improvements with no formal arrangements or contracts in advance.[84] Ray has noted that Pan Am worked in a similar way when establishing the Takoradi Route, but it is important to note that such a system likely worked smoothly in British colonies because the British and Americans were generally trusting of each other and had mutually beneficial goals. Such was not the case with the "Free" or "Fighting" French of Equatorial Africa, with whom a number of rather complicated problems arose.

First there was the issue of the ambiguity of the Free French role during the early part of the war. As the leader of this resistance movement (including its formal organization The French National Committee, based in London), De Gaulle was anxious to obtain formal recognition from Britain and the United States as a government in exile and member of the Allied Front. He was also aware that the territories of French Equatorial Africa were vital to the allied resupply effort. Thus, in June of 1941—one month after Petain's announcement of the Paris Protocols (intended to grant the Germans use of French military facilities in Syria, Tunisia and French West Africa)—De Gualle sent his representative Rene Pleven to Washington in hopes that by offering the use of airports in his territories he could convince the U.S. government to officially recognize the Free French.[85] Roosevelt declined to meet with him; however, at the time official U.S. policy was to work with the Vichy government in an attempt to convince it not to meet all of Hitler's demands. Out of fear of offending the Vichy regime, the State Department did not officially receive Free French representatives until October of that year.

This political slight exacerbated deep feelings of mistrust among the colonials in French Equatorial Africa, who were dealing with a number of other issues as well. Among these were the shame of being occupied by Germany and having to submit to American authority, uncertainties concerning what the nature and constitution of the French government would be after the war and worries about how they would sustain their power in nations increasingly demanding liberation from under the colonial yoke.[86] These problems led to a defiant attitude toward, and at times rather acrimonious relations with, U.S. government representatives. These dynamics played into the early episode of oil history at Pointe-Noire documented below. The data provided illustrate not only how the international oil companies sought to interpose themselves between these various and contending forces to keep their businesses profitable, but also how the U.S. government both aided the companies and assuaged French hostility through the manipulation of contracts and strategic apportioning of Lend-Lease Aid.

This episode begins with a confidential intelligence report on the subject of "French Equatorial Africa: Air Fields and Seaplane Facilities" that was issued by the U.S. Chief of Naval Operations on July 10, 1941.[87] Compiled within a month of Pleven's Washington visit and based on conversations with him, the report provided information on all "outstanding landing fields and possible seaplane harbors" in the colony, but judged the facilities at Pointe-Noire, Middle Congo, to be most suitable for American use. As a result of this report, the U.S. government put in a formal request for the French National Committee to "study favorably" the use of the airdrome at Pointe-Noire and take steps for the immediate undertaking of improvements. These included lengthening the landing strips, installing radios, building hangars and lodging facilities and installing large fuel storage facilities and pipelines at the port.

Little activity in this direction appears to have taken place until February of 1942, when Adrien Tixier, a member of the Free French Delegation in Washington, requested a meeting with the Assistant Secretary of State Adolph A. Berle.[88] The purpose of the meeting was to inform Berle as to the National Committee's decision on the issue. Tixier began by attempting to gain the same concessions Pleven had, stating that although it was entirely within the policy of the French National Committee to grant use of the facilities at Pointe-Noire, "the French National Committee felt that if the United States requested the collaboration of Free France, it ought to find some way of recognizing Free France as an ally."[89] Tixier additionally indicated that the Committee desired the provisioning of eight Lockheed Lodestar transport planes as part of the negotiations, these being necessary because of the decrepit state of existing planes and the fact that similar requests had been put in to the British with no results.[90] Berle's response was one that he would reiterate many times over the next six months: whereas he was willing to look into the possibility of supplying the transport planes through the Lend-Lease program, their provision could not be used as a bargaining point in the formal agreement for the use of Pointe-Noire.[91] This response apparently did not sit well with the National Committee, for within a month the War Department in Washington was informed by General Valin, De Gaulle's air chief, that he "fears disagreeable incidents will occur if American aircraft or personnel arrive at Pointe-Noire before agreement is reached on [the] Fighting French request for eight liaison aircraft."[92]

By July of 1942 it appears that the United States had tired of waiting for an official clearance from the National Committee, for in a Memorandum sent by the Division of Near Eastern Affairs an unknown author remarks (regarding the commencement of improvements at Pointe-Noire)

> I discussed this matter with Mr. Berle and we agreed that in view of the extreme urgency of this construction, particularly in view of the reverses in North Africa, the War Department should go forth with this construction by agreement with local authorities and that the Department of State should support the War Department in this matter if the Free French Committee in London interpose objections. We agreed that the memorandum now being discussed with General de Gaulle might take care of their feelings over this matter but that in any event the State Department should endeavor to "pick up the pieces" if the War Department's proceedings on this necessary construction caused further difficulties.[93]

The U.S. military had in fact entered into a verbal arrangement to begin improvements as early as April of 1942, working through the local French Military Commander General Sice.[94] His willingness to work with the Americans, especially without pushing for U.S. recognition or the delivery of the transport planes, greatly angered the National Committee, which censured him by removing him from office.[95] The Committee, however, was left in an awkward position because work had already begun; as a result they decided not to press the conditions and allowed the work to continue. Future problems of this type were avoided by requiring that the National Committee approve each stage of the improvements, such as the construction of the runway, the erection of the gasoline reservoirs, the laying of the pipeline and the construction of a hotel for Americans.

On July 20, 1942 Adrien Tixier again approached the Assistant Secretary of State, this time regarding the bulk storage and handling facilities that were being erected in the port.[96] The U.S. Military contracted the job out to Socony-Vacuum, who had one week earlier brought in one hundred U.S. oilmen to begin work on the oil tanks and pipelines.[97] Tixier reported to Berle that the National Committee was "anxious to know whether, at the end of the war, the permanent installations at Pointe-Noire would revert to France without payment or indemnity of costs."[98] From the ongoing waffling at the State Department regarding this request (to be documented below), it appears that post-war disposition of facilities was not an issue they had taken into account. Berle, for example, first told Tixier that his question would be best posed to the U.S. military; subsequently, however, he stated that in his personal opinion, "the installations should revert to France because it would cost more to tear down houses, hangars, radio masts, etc and transport them back to the U.S."[99]

Likely to have provoked the National Committee's question was the fact that they were in the process of making arrangements with the British for the construction of oil reservoirs to fuel British ships at the same time.[100] The Admiralty contracted the job out to Royal Dutch Shell, and the supplying of fuel was to be for military ships alone. According to the agreement, the British would build the facilities at their expense, and the French would operate them and make them available to the British without a fee for the duration of the war. When the war was over, however, they were to become property of the French. Nearly a year later Lawrence Taylor, the American consul at Brazzaville (the capital of Middle Congo) learned that the Free French had pushed an exceedingly hard deal with the British; in a move that was unusual for the time, the French had refused to sell Royal Dutch Shell the land where the installations were to be built, stating that they "disapproved of the advantage this would give one company over another."[101]

By the end of August the Americans had agreed on a similar deal with the Free French; the facilities would be handed over to the French upon the war's end, but the U.S. Army was allowed to remove any moveable property if they desired to do so.[102] In the negations surrounding this agreement the Free French were again denied their request of eight transport planes, this time on the grounds that "the completion of the Pointe-Noire project is of common interest in the war effort and should be dealt with strictly on its own merits." The Free French actions in these deals indicate that beyond seeking access to military supplies, they were keen to secure lasting benefits (regarding fuel storage and transportation facilities) for the colony after the war as well. The French government had been prospecting in the colony (near Pointe-Noire, Libreville and Doula) since 1935, and promising fields had been located at the latter two locations. This work had been carried out by the *Companie Francaise de Petrolieres* or (CFP; est. 1924), and all prospecting rights belonged to the colonial government. Due to the outbreak of war, however, further developing of the fields was halted.[103]

In October 30, 1942, the Assistant Secretary of State in Washington (Berle) reported that he had been approached by Tixier with another question regarding the oil installations at Pointe-Noire.[104] It emerged that Harold Sheets, the Vice-President of Socony Vacuum, had met with Tixier and asked him for permission to lease the very same facilities Socony was constructing at Pointe-Noire. Tixier was surprised by this request, and approached Berle to clarify the situation. Berle, in turn, consulted Colonel Marlowe M. Merrick of the Army Air Force, who reported that Sheets was not acting on the part of the military but on behalf of his company alone. He informed Berle that a large oil field had been discovered close to the coast ("in the back of Loango and Pointe-Noire) and that Royal Dutch Shell was currently carrying out subsoil exploration. Although a French company held the concession for the field, they were, according to Merrick, "in no position to develop," and as a result Royal Dutch Shell and Socony-Vaccuum had entered into partnership with the firm. Merrick noted that "pipelines from the field to the Pointe-Noire and other installations could be run easily," and that "oil could actually be obtained and laid down there within six months." Thus, the oil installations in question would "complement the field and make it very valuable."[105]

It must be noted that by the time Sheets approached Tixier about leasing the Pointe-Noire facilities (October 1942) West Africa's strategic role in supplying the Allies had begun to decline.[106] The impact of this shift can be seen in subsequent U.S. government attempts to alter the terms of the deal, rising suspicions between the various foreign actors in Pointe-Noire and an increasing frustration on the part of the American Ambassador Lawrence Taylor regarding Free French demands for continued Lend-Lease Aid. For example, between the months of February and April of the next year (1943), the U.S. consulate established an agreement with the Free French regarding the use of facilities in Ft. Lamy, Chad. In putting together the agreement, Taylor asked for the opinion of the International Petroleum Committee regarding the post-war disposition of fueling facilities at Ft. Lamy, which constituted the major part of the installation. The resultant agreement differed from previous ones in that it allowed for future removal of the installation (by the U.S. Army) "without qualification as to time and circumstance."[107]

This new agreement was also used as a springboard for entering into renegotiations regarding the facilities at Pointe-Noire. As Paul Alling (Division of Near Eastern Affairs) noted " . . . the agreement signed with the Fighting French regarding the Pointe-Noire installation called for the reversion of the whole installation without cost to the French State at the end of the war. The object of our re-opening negotiations was to set aside this requirement and obtain other terms."[108] Accordingly, a new agreement was drawn up by Lawrence Taylor and submitted to the Secretary of State on April 12, 1943. As was the case with the first agreement, it stipulated that the Pointe-Noire installation would be constructed, managed and leased to a contractor that the U.S. government selected and would remain property of the U.S. government until the war's end. Dramatically different, however, and most likely engendered by Socony's interest in the oil field discovered near Pointe-Noire, was the stipulation that "the contractor shall have the right to sub-lease any or all installations upon thirty days notification in writing," and that "final disposition of all installations will be the subject of further negotiation between the two countries."[109]

These attempts to alter the terms of earlier agreements—agreements that had been made during the most desperate months of the Allied war effort—reveal that in terms of its commercial relations with Africa, the U.S. Government had also begun to look to the future. Whereas protecting the rights of American businesses to operate in colonial markets was a key element of U.S. foreign policy toward Africa in the post-war years, this example illustrates the ways that the Lend-Lease program served as an important early tool toward this end. It is interesting to note, however, that Consul Taylor was not privy to the joint agreements between American, British, and French oil companies until well after they were established and that he learned of these through the U.S. military rather than the companies themselves. This may have been due to the fact that much of Socony-Vacuum's work in the seaport was being paid for through Lend-Lease.

Apparently concerned about this very issue with regards to the British, Taylor had sent a telegram to the Secretary of State in March of 1943 stating that "in spite of the lessened importance of French Equatorial Africa from a military point of view, the British are continuing with the installation at Pointe-Noire of storage facilities for ship's fuel oil."[110] He noted that the improvements were quite substantial (two tanks of two million gallons capacity and another of 300,000-gallon capacity for diesel fuel, and a pipeline with six outlets that could discharge 90,000 gallons of oil per hour) and was apparently so disquieted by the possibility that the work was being paid for through Lend-Lease that he asked the Naval Division Office to investigate how the British were obtaining their supplies.[111] The Naval Division reported that Lend-Lease supplies were not being used; Shell had purchased all of them through a British steel company that had a subsidiary office in Cape Town.[112]

Taylor became increasingly focused on what he considered abuse of the Lend-Lease system as the months wore on. He was particularly obsessed with the requests of the Free French. In May of 1943 he wrote to the Secretary of State asking for an "adjustment to conditions," noting that since the success of the 8th army Libyan campaign, the strategic value of the region was very small" and that "to send here now equipment urgently needed seven months ago is wasteful."[113] He had put together a report on the problem in April of 1943, and provided a number of examples of the way the system was being manipulated by the Free French. "The Colony considers itself the 'Cinderella of the The French Empire,'" he stated, "for it has long been neglected by France. There is already evidence that an effort will be made to stretch the Lend-Lease program to cover belated improvements, comforts, and luxuries to which the colony has never been accustomed."[114]

As one example he cited the fact that although there had formerly been 10,000 troops in the colony, they were all now gone. Why then, he asked, was the French administration asking for more military supplies? According to him the administration had begun hastily recruiting African soldiers (who under regulation could not carry arms until the completion of three years' training) simply to be able to order military equipment under Lend-Lease. Another case was the request for a diesel tractor for a timber project in Gabon. Upon investigation, it was revealed that the tractor was designated for work in building a road that connected a small gold mine to two timber reserves that had to rely on head porterage to obtain supplies.[115] "Since neither gold nor timber from the area were needed in the war effort," he considered the project to be an example of how the Free French were attempting to exploit the system.[116] He had hoped that "goodwill" would eliminate this problem but noted that "instead of being grateful to the British and the Americans for sharing their scarce materials, local French tend to look upon Anglo-American activity with a great deal of animosity. The thought in local administrative circles is that the British and Americans are prying too actively into French affairs."

Interesting to this case is the fact that Taylor, despite his position as the American Consul at Brazzaville, appears to have been largely out of the loop regarding both oil and Lend-Lease deals. The latter is evidence by a telegram he sent to Henry S. Villard, Assistant Chief of Near Eastern Affairs in Washington in May of 1943. After thanking him for information regarding Lend-Lease in a previous telegram, he states

> The confusion of my previous telegrams and dispatches regarding Lend-Lease is due to my ignorance of my operation of the law. I was never informed that an agreement was made with the Fighting French. I learned it by accident from an American Army Officer who was passing through and who had a copy of the "Sixth Report." I would be pleased to have anything in addition with [*sic*] tells how the system works. The Reports to Congress give only results and are mostly propaganda.[117]

Despite the conciliatory tone, Taylor's frustration with the situation is made evident in his last line: "I would like to know if this area is to be supplied with everything it asks for, or if it is intended that total shipment here from Great Britain and the United States are to be held down to reasonable maxium [*sic*], and how it is to be accomplished."[118] By August of 1943 Taylor's anger was even more palpable, made evident in a letter to the Secretary of State that exhibits a degree of impertinence rarely seen in State Department documents.[119] In describing his interactions with the Lend-Lease office in Washington, he noted that

> The naïve phrase of one Lend-Lease telegram struck me funny. "If they do not need it, why do they order it?" I suppose it is hard for some people to realize that this area has been petted and spoiled by British indulgence, that is [sic] has no conception of war sacrifice, and is living in an abundance it has never known before the war. Lend-Lease is considered a grab-bag out of which you get all you can if not for the present then for the future. It is for that reason that a control must be established at this end in order to sift out the requirements for the continuation of normal activities as compared with new constructions or unnecessary luxuries.[120]

Although the Division of Near Eastern Affairs forwarded Taylor's May 7 report to the State Department and asked for action to be taken along the lines he suggested, it appears that there was no official response until March of the next year (1944), when a document referred to as "Williamson Report on French Equatorial Africa" was circulated in official diplomatic circles.

Whereas the report has not yet been located in State Department files, it appears to have been an exposé of the activities of the Free French in relation to Lend-Lease.[121] In a clear attempt to downplay its effects, the State Department described the report as having "aroused an undue amount of surprise and indignation, and given rise to moral considerations which are rather beside the point."[122]

The memorandum regarding the Williamson Report is rather unusual in the degree to which it strives to absolve the Free French of responsibility for any improprieties. Whereas the State Department acknowledged that the British had allowed the Free French to order supplies that had little or nothing to do with the war effort, it argued that "if concrete reinforcing bars arrived after the need for military constructions had disappeared, it is not extraordinary that they should be used even for non-military purposed rather than be allowed to rot."[123] Furthermore, one could not blame the French, for according to the State Department, the burden of proof regarding the legitimacy of Lend-Lease orders was on the supplying countries. It was because they had never been given any instructions that the problems had arisen; if such had been made clear, they argued, many of the problems in the Williamson report would never had arisen. The matter was put to rest in the last paragraph of the memorandum, which stated

> In conclusion, such fault as there has been lies as much with the British and ourselves as with the officials of FEA. I can find no evidence in this report of our ever having made our position unequivocally

> clear as to what we expected to obtain from the areas in question to further our general war effort, what we intended to put in other areas for the purpose or the methods we considered proper for the distribution and use of the material provided. It is naïve and futile to expect that a backward impoverished colonial country will exert the same controls as we ourselves do, unless we make the rules of the game clear beyond the possibility of doubt and provide in advance for their adequate enforcement.[124]

For those aware of the history of post-independence in Africa, when African rulers carried out similar improprieties with Western-supplied aid, the logic of the State Department's argument is rather telling: in poor, underdeveloped and politically 'backward' countries, it is to be expected that rulers will appropriate aid for their own benefit; if such action is beneficial to American interests, the U.S. government will turn a blind eye. It is rather surprising, however, that the first case of such improprieties involved the Free French and that during this early era of aid to Africa, the "donors" were deemed to be responsible for the corruption that took place.

Further research is required to fully reconstruct this episode of history; key to an understanding of the events would be the locating of either the Williamson Report or the "Sixth Report" referred to by Taylor. It seems clear, however, that the U.S. government at times used the Lend-Lease program to further both American and American oil companies' interests. This, however, is not entirely groundbreaking data, for similar uses of Lend-Lease are documented for Liberia and Saudi Arabia.[125] Perhaps more important is the evidence to suggest that the U.S. State Department, despite its willingness to institute policy beneficial to the oil companies, was never fully apprised as to the companies' actual operations and business strategies in the field. This is a phenomenon that, as is now documented here, began during the colonial era. It continued well into the era of African independence and oil discoveries (as my earlier works have shown),[126] and as I will argue in later works, played an important role in contributing to the development of the "Oil Curse" in sub-Saharan Africa.

CONCLUSION

This study has gone a long way toward "inserting" Africa into the global histories of oil that have for the most part neglected it. By reconstructing the early history of Socony-Vacuum in Africa, especially as pertains to its expansion into markets across the continent, we are provided with a framework for the reconstruction of more people-centric histories that chronicle the impact of petroleum products on society at large. This background is also essential to understanding how the Takoradi Route was established so quickly and was so successful in helping to gain victory over the Axis

Powers in World War II. Whereas all previous studies have focused on the role of Pan-American Airways in developing this route (and are of course fascinating in their own right), this article has gone further to suggest that it was American oil companies that led not only Pan Am in these efforts but also the United States Government, as it made its first foray into the African geographical, political and economic terrain. Furthermore, this chapter argues that it was through the Lend-Lease program that the United States first began to formulate its policies for aid and commerce in Africa. Through this experience, a number of practices that would prove detrimental to the proper working of both the newly independent African governments and the newly burgeoning oil industry were set in place. Among these was the government's tendency to promote U.S. oil companies' interests while remaining rather uninformed about (the latter's) actual operating and business strategies, as well as turning a blind-eye to corruption in underdeveloped countries as long as it served U.S. political interests.

The study has shown that the history of the "downstream" (i.e., marketing and distribution) sector is as important to study as the "upstream" (exploration and manufacture). The next task (for myself and/or other historians of oil in Africa) is to understand how practices and procedures associated with this earlier phase of oil history in Africa were continued, adjusted or jettisoned as the era of oil production began.

NOTES

1. Research for this article was made possible through the generosity of several funding sources at the University of Houston between 2005 and 2009: The African-American Studies Program (Summer Research Travel Grant); The Department of History (two travel and research grants); The Division of Research (Small Grants Program and Grant to Enhance and Advance Research [GEAR]). I would like to thank numerous members of the Department of History, both faculty and graduate students alike, for their thoughtful comments on the paper and ongoing guidance on the larger project. Key among these are Tyler Priest, Joe Pratt, Nancy Beck-Young, Tom O'Brien, Bob Buzzanco, Jason Theriot and Jaime Christy. Equally important have been the comments of Adam Paddock, co-editor of this volume, and the contributions of four very capable research assistants: Tomiko Meeks-Henderson, Anna Fields, Kinal Patel and Samantha Rodriguez.
2. Author unknown. "The Saga of the Seven Engineers," *Socony-Vacuum News,* August 1944, 14. Exxon-Mobil Archives, University of Texas, Austin, "Africa General File," 2.207/E165.
3. Ibid. These were: Socony-Vacuum, Texas, Ethyl Corporation, Standard Oil California, Phillips Petroleum, Standard Oil New Jersey and Sun Oil.
4. Deborah Wing Ray, "The Takoradi Route: Roosevelt's Prewar Venture Beyond the Western Hemisphere," *The Journal of American History* 62, no. 2 (1975), 342, 346.
5. Ibid., 344.
6. Tom Culbert and Andy Dawson, *Across the Sahara in 1941 with Pan Am* (Paladwr Press, 1998); Deborah Wing Ray, "The Takoradi Route," 340–358;

John Schwally, *FDR's Secret Air Force* (JFS Films, New York) (forthcoming; preview trailer can be viewed at http://www.jfsfilms.com/).

7. Ibid., 355.
8. Ibid.
9. These sources include company documents (mainly advertising, publicity and company journals) from the Mobil-Exxon archives at the University of Texas, Austin; State Department documents from Record Group (RG) 59 and 89 of the National Archives and Records Administration (NARA) in College Park, Maryland; articles from *The West African Review,* a colonial journal published by the Elder Dempster Shipping Line; and numerous secondary sources relating to oil history, oil company histories, African history and World War II.
10. Standard Oil of New Jersey, or more commonly, "Jersey" or "Jersey Standard"; Standard Oil of New York (SOCONY); Standard Oil of Ohio (SOHIO); Standard Oil of Pennsylvania. A great many more standard companies were to be created before the antitrust break up of 1911 (e.g., Standard oil of California). These, however, were the original four. Jersey Standard eventually became Esso, and later Exxon; Standard Oil of New York would eventually become Mobil, now part of ExxonMobil; Standard Oil of Ohio became SOHIO, now part of BP.
11. Daniel Yergin, *The Prize: The Epic Quest for Oil, Money, and Power* (New York: Simon and Schuster, 1991), 43, 51. For a full history Standard Oil's early years, see Yergin, Chapter 2 and Ralph Hidy and Muriel E.Hidy, *The History of Standard Oil Company (New Jersey):Pioneering in Big Business, 1882–1911* (New York: Ayer Books, 1987), Chapters 2 and 3.
12. Anthony Sampson, *The Seven Sisters: The Great Oil Companies and the World They Shaped* (New York: Bantam Books, 1975), 34–35.
13. Hidy and Hidy, *History of Standard Oil,* 54.
14. Ibid., 42.
15. Ibid., 485.
16. Ibid., 486.
17. Ibid., 485.
18. Ibid.
19. Ibid., 122–123.
20. "Vacuum Oil Company–Egypt" (schedule of managers and first leases taken), ExxonMobil Archives 2.207/E171, "Sudan Files, Folder 1." This document indicates that the Vacuum Oil Company Trade Mark was registered in Egypt in 1896, but that land was first leased for an affiliate office in 1906.
21. "What is this Vacuum Oil Company?" (Article from the *Mobil Monitor* [South Africa], Christmas 1981). ExxonMobil Archives 2.207/E169, "Africa: South Africa General, 1911, 1959–89."
22. Ibid.
23. Ibid.
24. "Being There When All Began" (article in *Mobil Monitor* [South Africa], Christmas 1980), ExxonMobil Archives, 2.207/E169, "Africa: South Africa General, 1911, 1959–89."
25. "What is this Vacuum Oil Company?"
26. "The Origins of the Flying Red Horse" (typed manuscript attached to letter to C.E. Solomon in New York from S. J. Wilson in Cape Town, Feb. 13, 1963). ExxonMobil Archives 2.207/E169, "Africa: South Africa General, 1911, 1959–89."
27. No title (Full page advertisement from *The Gargoyle*—a magazine for motorists published by Vacuum oil in South Africa, 1913). Exxon Mobil Archives, 2.207/E169, "Africa: South Africa General, 1911, 1959–89." Note: it is

possible that this advertisement is from a later date, despite the fact that the archivist at Vacuum Oil who filed it noted a 1913 date on it. This is because subsequent documents indicate that the first edition of The Gargoyle was published in 1924 (see "Rich Storehouse of Company Lore," article in *The Mobil Monitor,* undated [but post-1981], Exxon Mobil Archives, 2.207/E169, "Africa: South Africa General, 1911, 1959–89").

28. Ian Cummins and John Beasant, *Shell Shock: the Secrets and Spin of an Oil Giant* (Edinburgh, Scotland: Mainstream Publications, 2005).
29. For these early stages of Shell oil and its challenge to Standard in the 1890s, see Daniel Yergin, *The Prize: The Epic Quest for Oil, Money, and Power* (New York: Simon and Schuster, 1991), Chapter 3. Not having done research in Shell's company archives, it is rather difficult to reconstruct the early marketing history of Shell in Africa. Although Jonker and Luiten van Zanden mention that Royal Dutch/Shell had, in terms of marketing, "control over the east coast of Africa" in 1919, 267. Africa is not otherwise mentioned in early histories of the company before the 1928 Consolidated and Achnacarry Agreements. Shell entered into these partly as way of gaining better access to African and Indian markets (see discussion below). See Joost Jonker and Jan Luiten van Zanden, *A History of Royal Dutch Shell, Volume 1: From Challenger to Joint Industry Leader, 1890–1939* (Oxford: Oxford University Press, 2007), 278–283; for the history of these agreements from Shell's perspective, and James Bamberg, *British Petroleum and Global Oil, 1950–1975: The Challenge of Nationalism* (Cambridge: Cambridge University Press, 2000), 252–257 for the BP perspective.
30. "Vacuum Oil Company–Egypt" (schedule of managers and first leases taken).
31. Ibid.
32. "Delivering Gargoyle Products in the Sahara," (photograph with caption, *The Gargoyle,* 13 January 1915), ExxonMobil Archives, Egypt File #2 (2.207/E166).
33. Letter from Socony-Vacuum manager in New York (William N. Mann?) to newly hired salesman (John Blank?), December 31, 1909, ExxonMobil Archives 2.207/I166 "Middle East #2, Algeria Folder."
34. Hidy and Hidy, *History of Standard Oil,* 490.
35. "In the Levant" (half-page advertisement, *Vacuum Oil News,* March/April 1927), ExxonMobil Archives 2.207/E171 "Sudan Folder." This prideful emphasis on the employment of local peoples, a phenomenon that becomes increasingly common in literature produced by petroleum distributors in Africa during the first half of the century, provides an interesting counter-example to the resistance toward the indigenization of the labor force once production begins to take place (1960s).
36. Ibid.
37. "Vacuum Oil Company–Egypt" (schedule of managers and first leases taken).
38. "A Gargoyle Correspondence School" (article from *The Vacuum Oil News,* November 1927), ExxonMobil Archives, 2.207/E166, "Egypt File #2."
39. "Socony-Vaccum Oil Company, Inc.–West Africa: Historical Background" (typed manuscript marked "confidential" with no date, but latest data from years 1951–53), ExxonMobil Archives 2.207/E165 "General West Africa '27–43, '52-'70 File."
40. "Mobil Oil Ghana Limited" (print article/pages from company print document [handbook?]), ExxonMobil Archives 2.207/E165 "Ghana File #1."
41. "West Africa Booklet," 2, Appendix B, "Training" (January 1958), ExxonMobil Files 2.207/E165 "Africa General File."

42. Ibid.
43. Booklet, "General Information About Mobil Overseas West Africa Interests" (Company Handbook, January 31, 1956), ExxonMobil Archives, 2.207/E165 "Africa General File–WA 1958–65").
44. "Tut! Tut!" (article in *The Sales Generator,* June 1923), ExxonMobil Files, 2.207/E166, "Egypt File #2."
45. Ibid.
46. "In West Africa: A Thumb-Nail Picture of One of the Company's Marketing Territories," 6.
47. Author unknown, "British West African Shipping," *West African Review,* London (April 1937), 97. As such, ships carrying vegetable or seed oils were often targeted by German forces for destruction.
48. "In West Africa: A Thumb-Nail Picture of One of the Company's Marketing Territories," 7.
49. Ibid.
50. George P. Kerr, *Time's Forelock: A Record of Shell's Contributions to Aviation in the Second World War* (London: Shell Petroleum, 1948), Introduction.
51. Ray, *The Takoradi Route,* 341–342.
52. As was noted above, see See Joost Jonker and Jan Luiten van Zanden, *A History of Royal Dutch Shell, Volume 1: From Challenger to Joint Industry Leader, 1890–1939* (Oxford: Oxford University Press, 2007), 278–283 and James Bamberg, *British Petroleum and Global Oil, 1950–1975: The Challenge of Nationalism* (Cambridge: Cambridge University Press, 2000), 252–257. Also, from the perspective of Burmah Oil, T. A. B. Corley, *The History of the Burmah Oil Company: Volume II, 1924–1966* (London: Heinemann, 1988), 30–31.
53. T. A. B. Corley, *Burmah Oil Company,* 18.
54. Yergin, *The Prize,* 262.
55. Ibid., 263.
56. Bamberg, *British Petroleum and Global Oil,* 252.
57. Ibid., 253. As Bamberg indicates, however, the terms of the marketing arrangements were in reality quite unequal because Shell was given the job of managing Consolidated Oil and Anglo-Persian completely ceded control over both its brand and trademark within the region as part of the deal. This became a serious point of contention between the two companies after World War II, when Anglo-Persian (now called BP) acquired the Atlantic Richfield Company and its African subsidiary, Atlantic Richfield Refiners of Africa. Because the latter marketed products across southern Africa, and South Africa itself was by this time the "plum" of the Consolidated Area (with the largest and most profitable market in the region), BP began to chafe at the restrictions imposed by the 1928 agreement. It was not completely dissolved, however, until 1975. For this latter history see Bamberg, *British Petroleum and Global Oil,* 260–263.
58. Jonker and Luiten van Zanden, *A History of Royal Dutch Shell,* 79; Bamberg, *British Petroleum,* 255.
59. Jonker and Luiten van Zanden, *A History of Royal Dutch Shell,* 191.
60. Yergin, *The Prize*, 263.
61. Ibid.
62. Ibid., 260.
63. Ibid., 264. At its core was an agreement that each company would be allocated a quota in various markets, that they would be paid a percentage of the share of total sales based on the gains their companies had made in 1928, that volumes could be increased only so far as demand grew and that they

would always retain the same percentage share. The companies also made plans to share facilities and be more cautious in building new refineries and other facilities, a way to avoid duplicating each other's efforts and reduce costs. Also key was the agreement that each market was to be supplied from the nearest source, thus saving transport costs. Within a few months of the meetings the companies also agreed to control their production; if volumes were increased in this domain, they were obliged to sell their product exclusively to other members of the "pool."

64. Ibid., 267.
65. See "West Africa's Place in Imperial Defense," *West African Review*, November (1937): 16, for information on the buildup of storage facilities. That it was RD/Shell supplying them is made evident in "Socony World: The Strategy of Oil," *Fortune Magazine* 7, February (1943): 177.
66. Ibid.
67. "The Empire and Beyond" *West African Review*, Nov. 30 (1937): 20.
68. Kerr, *Times Forelock*, Chapter 1.
69. Ray, *Takoradi Route*, 342.
70. "Socony World."
71. As Ray has chronicled, this modernization program was referred to as the Airport Development Program or ADP. It was paid for by War Dept funds, although "the public was encouraged to assume that work on the airfields was being done by PAA for its own purposes," *Takoradi Route*, 343.
72. Ibid., 356.
73. Ibid., 343.
74. Ibid., 342.
75. "West Africa Booklet, First Edition, 1953" (Company Handbook), ExxonMobil Archives 2.207/E165 "Africa General: West Africa, 1946–56."
76. Ray, *Takoradi Route*, 343.
77. "Strategy of Oil," 80.
78. "Fueling Warplanes in Darkest Africa: Socony-Vacuum Service Vital to U.S. Ferrying Command" (article in *Socony-Vacuum News*, October 1942), ExxonMobil Archives 2.207/E165, "African General" File.
79. This was done by forcing large wooden slats under the wheels of the trucks, which allowed for better traction.
80. "The Strategy of Oil," 116.
81. Ashely Jackson, *The British Empire and the Second World War* (London: Hambeldon and London, 2006), 217.
82. This elimination of local retailers was a phenomenon that affected nearly all businesses; as an article in the *West African Review* of January 1943 noted, "Government control has been extended to cover most commodities. Merchants who formerly bought and shipped for sale through brokers now act as buying agents for the ministry of supply, whilst the brokers have been almost entirely eliminated from the trade. Many of these are old established firms who have formed a vital part in the machinery which made London the chief center of the trade in colonial produce," "West African Trade Leaders Review 1942," *West African Review*, January (1943): 61. It is not yet clear if this system applied to all petroleum products in all African colonies, for it appears that in French Equatorial Africa the oil companies continued to supply lubricating oils to local commercial agents until May of 1943. At that point the National Committee (leadership of the Free French) insisted that all such products be supplied through the Lend-Lease program. This move was opposed by the American administration and the oil companies ("wholesalers) alike. As Lawrence Taylor, the American consul at Brazzaville noted, "Everyone is satisfied with the present arrangements, whereas if the oil is sent

under Lend-Lease the Administration here will have to receive the supplies, control the distribution, and render the accounting. The wholesalers will lose their commission and are therefore also fighting the change." "The Application of Lend-Lease in French Equatorial Africa," telegram to the Secretary of State from Taylor, May 7, 1943. NARA RG 59, "French Equatorial Africa 1940–1963," Box 5225. More research is required to understand the full implications of this move.

83. "Mobil Operations in Africa," internal report, October (1961), ExxonMobil Archives 2.207/E165 "Nigeria" Folder.
84. The problems inherent will be discussed at length below.
85. After the German occupation (May 10 to June 22, 1940) and de Gaulle's famous "Appeal of July 18th," Cameroon and all eventually all the colonies of French Equatorial Africa chose to side with the Allies, as opposed to the colonies of French West Africa, which remained tied to the Vichy Government in France.
86. A rather poignant portrayal of the mentality of the Fighting French is provided in a telegram sent on April 7, 1943, to Henry S. Villard, Assistant Chief of the Division of Near Eastern Affairs in Washington by Lawrence Taylor, American Consul at Brazzaville: "This had been an extremely interesting post. The problems here have been new and novel. The psychology of the French people here, regarded as heroes by themselves and as rebels by others in [*sic*] an interesting study. Their disappointment at not finding themselves the leaders of a great overwhelming movement and their fear of being misunderstood when it is all over are both very real. They indulge in noble thoughts and self-pity by turns. They feel themselves adrift on a sea, with the beacon light which has been guiding them growing fainter rather than brighter as time goes on." NARA RG 59, "French Equatorial Africa 1940–63" Box 5225. As will be documented below, his attitude toward the Fighting French becomes much more negative and cynical as the war continues on.
87. NARA RG 59, "French Equatorial Africa 1940–1963," Box 5225, Folder 4.
88. "Memorandum of Conversation–Airdrome at Pointe Noire; Need of Air Transport in French Africa, July 10, 1941." NARA RG 59, "French Equatorial Africa 1940–1963," Box 5225, Folder 4.
89. Ibid. Note that these are Berle's words (reporting what Tixier said).
90. Ibid.
91. This issue is brought up six times between February and August of 1942. The full (translated) text of a letter from de Gaulle to the U.S. military commander in charge (General Bolte) arguing for the supplying of these aircraft can be seen in a telegram from "Winant" (title unknown) to the Department of State on July 7, 1942. NARA RG 59, "French Equatorial Africa 1940–1963," Box 5225, Folder 4.
92. Telegram from "Hull" (title unknown) to State Department, August 21, 1942. NARA RG 59, "French Equatorial Africa 1940–1963," Box 5225, Folder 4.
93. "Memorandum" from Department of State, Division of European Affairs (sender and recipient both unclear), July 3, 1942. NARA RG 59, "French Equatorial Africa, 1940–1963," Box 5225.
94. In a letter dated April 2, 1942, to Patrick Mallon, the American Consul at the time, Sice stated: "I have the honor to conform by means of this letter that we will increase our efforts tenfold in order to make the airfield at Pointe-Noire entirely practicable for American planes with the shortest delay. You may count on us and guarantee to your government that the airfield at Pointe-Noire the use of which was suggested last year to President Roosevelt's government by General de Gaulle, will be in condition to give adequate service." Quoted in "Agreement with Fighting French Regarding

American Installations at Pointe Noire" (Telegram sent by Lawrence Taylor, American Consul, to Department of State, 7–27–42). NARA RG59, "French Equatorial Africa, 1940–1963" Box 5225.

95. Ibid.
96. "Memorandum of Conversation" (between M. Tixier and A. A. Berle, July 20, 1942), RG 59 "French Equatorial Africa, 1940–1963," Box 5225, Folder 4.
97. "Public Relations at Pointe-Noire" (telegram from Lawrence Taylor, American Consul at Brazzaville to Secretary of State, 7–14–42). NARA RG 59, "French Equatorial Africa 1940–63," Box 5225, Folder 4.
98. "Memorandum of Conversation" (between M. Tixie and A. A. Berle, July 20, 1942).
99. Ibid.
100. "Agreement with Fighting French Regarding American Installations at Pointe Noire" (Telegram sent by Lawrence Taylor, American Consul, to Department of State, 7–27–42).
101. Letter from George A. Gade, U.S. Naval Liaison Office, Pointe-Noire, to Lawrence Taylor, American Consul, May 12, 1943. NARA RG 84, "French Equatorial Africa, 1942–1945," Box 6, Folder 7. Note that the quoted passage refers to Gade's words, perhaps not original wording used by Free French officials.
102. Telegram from Hull, Department of State, to General Dahlquist, American Embassy in London, 8–24–42. NARA RG 59, "French Equatorial Africa 1940–1963," Box 5225, Folder 4.
103. "Oil Resources in French Equatorial Africa and the Cameroons" (report sent by Lawrence Taylor, U.S. Consul, to Secretary of State, August 25, 1943). NARA RG 59, "French Equatorial Africa, 1940–1963," Box 5225.
104. "Memorandum of Conversation" (between Col. Marlow M. Merrick, Army Air Force, and A. A. Berle, Assistant Secretary of State," NARA RG 59, "French Equatorial Africa, 1940–1963," Box 5225 Folder 3.
105. Ibid.
106. As was noted above, the Takoradi Route was among the most travelled in the world between spring 1941 and summer 1942).
107. Letter from Paul H. Alling, Division of Near Eastern Affairs, to (unidentified) member of Aviation Committee, April 3, 1943. NARA RG 59, "French Equatorial Africa, 1940–1963," Box 5225, Folder 1.
108. Ibid.
109. "Report on Airfield at Pointe Noire for Hydrographic Office" (sent to Secretary of State by Lawrence Taylor, April 12, 1943). NARA RG 84, "French Equatorial Africa 1942–1945," Box 6, Folder 2.
110. "British Oil Installation at Pointe Noire" (telegram from Lawrence Taylor, American Consul, to Department of State, March 17, 1943). NARA RG59, "French Equatorial Africa, 1940–1963," Box 5225, Folder 3.
111. The response to this request (written by George Gade, U.S. Naval Division Office in Pointe-Noire, is included as an attachment to Taylor's telegram to the State Department informing it of the results of the inquiry. See "British Oil Instalation at Pointe-Noire" (telegram from Lawrence Taylor to Secretary of State, March 17, 1943). NARA RG 84, "French Equatorial Africa, 1942–1945," Box 6, Folder 2.
112. Ibid. (Gade's Telegram).
113. "The Application of Lend-Lease in French Equatorial Africa" (report sent by Taylor to State Department, May 7, 1943).
114. Ibid.
115. "Lend-Lease Supplies in French Equatorial Africa and Cameroon" (report prepared by Lawrence Taylor, sent to Division of Defense Materials, May 10, 1943). Included in "The Application of Lend-Lease in French Equatorial Africa" (report sent by Taylor to State Department, May 7, 1943).

116. Ibid. Interestingly, at the end of the report to the Division of Defense Materials and under the heading of "Strictly Confidential," Taylor suggested that a member of the French delegation was corrupt: "There is some uncertainty here regarding M. Levy of the Fighting French Delegation in New York. It is believed that he may attempt to make a make a personal profit out of the Lend-Lease shipments to this area. He is not regarded as being capable of judging accurately the requirements of the colony. "On the other hand full confidence has been expressed with regard to M. Nus." Whereas such may have been the case, one might suspect a degree of anti-Semitism at work as well.
117. Letter to Henry S. Villard, Assistant Chief Division of Near Eastern Affairs, from Lawrence Taylor, May 14, 1943.
118. Ibid. Note the similar focus on political use of Lend-Lease in a report submitted fourteen days later: "If it is a matter of policy to purchase the good will of the Fighting French by supplying material ordered for the Fort and Railroad, it is certain that those materials will be put to good purpose, that their use will result in the building and maintaining a port and Railway unequaled on the West Coast of Africa. . . . If, on the other hand, some consideration is to be given to the value of the contribution that can be made by this area to the war effort in repayment for the material sent, the schedule for 1944 should [b]e carefully scrutinized by persons qualified to do so." (Requirements for the Port of Pointe-Noire and Congo-Ocean Railroad report prepared by Lawrence Taylor and sent to the Department of State, May 28, 1943).
119. "Cooperation with Lend-Lease" (letter to Secretary of State from Lawrence Taylor, August 19, 1943). NARA RG 84, "French Equatorial Africa, 1942–1945," Box 6, Folder 2.
120. Ibid.
121. The first mention of this report appears in a State Department document entitled "Memorandum: United States Government, 'Williamson Report on French Equatorial Africa.'" NARA RG 59, "French Equatorial Africa, 1940–1963," Box 5225, Folder 2.
122. Ibid.
123. Ibid.
124. Ibid.
125. Ray documents the Liberian case, where the government, which "disliked the British and feared the Germans," opted to stay neutral in the war. Despite this Pan American, relying on a 1939 landing agreement with Liberia and funds from the War department, "hastily built a seaplane base in northern Liberia and a huge modern airfield just south of Monrovia." When Liberian officials learned of the military nature of these fields, "a generous outpouring of lend-lease funds quieted official doubts" (Ray, *Takoradi Route,* 354). The Saudi Arabian case is detailed by Robert Vitalis, *America's Kingdom: Mythmaking on the Saudi Oil Frontier* (New York: Verso, 2009), Chapter 3. In that case, according to Vitalis, the United States Government provided Arabian American oil company ARAMCO with Lend-Lease funds to build the kingdom's first major refinery, royal showcase farms, a railroad and a U.S. military base, the latter two being contracted in 1945 (after the effective end of the war).
126. "Oil Politics and Development in the Formation of A State: The Congolese Petroleum Wars, 1963–1968," *International Journal of African Historical Studies* 41, no. 2 (2008): 169–202; "Re-assessing the Role of Oil in the Nigerian Civil War," Paper presented at the 51st Annual Meeting of the African Studies Association, Chicago Il., Nov. 13–16, 2008.

2 Environmental Impact of Post-Colonial Mining Activities on the Jos Plateau

S. U. Fwatshak

INTRODUCTION

This chapter examines post-colonial mining activities on the Jos Plateau by focusing on two solid minerals, namely, tin (in former Jos Division) and Garnet (in Pankshin Local Government Council Area). It argues that after independence in 1960 tin mining continued to be practiced on a large scale using heavy machinery, which combined with colonial tin mining created huge environmental consequences. From the 1990s, tin mining was done on a small scale based on the use of low indigenous technology when imported heavy machinery could not be used owing to breakdowns resulting from old age and failure to service the machines with new parts or import new machinery. Indigenous technology led to minimal environmental effects on affected areas. It further argues that unlike the colonial period during which tin mining elicited serious reactions including episodes of violence, post-colonial tin mining did not elicit such serious reactions. With respect to garnet mining, the chapter argues that from its inception it was based on low-scale technology and yet created negative environmental effects and elicited reactions including episodic clashes between village land owners and miners. The environmental consequences of small-scale mining in both the tin and garnet areas include destruction to farm lands and public sources of water supply. Thus, whereas individual miners profited, the mined communities suffered various effects from mining operations.

Although tin and garnet are the main focus of this chapter, the Jos Plateau is endowed with various solid mineral resources but tin is the most famous and the most documented.[1] Of the numerous works three are outstanding. The first is Bill Freund's *Capital and Labour in the Nigerian Tin Mines*, C. K. Gonyok's, "A History of Labour on the Nigerian Tin Mines" and Nankan Bugudu's edited work, *Land Desecration and Environmental Degradation on Nigeria's Jos Plateau*. The first two are the most extensive and authoritative accounts of the tin industry and focus mainly on the production relations within the framework of international capitalism. There is some justification for the scholarly attention given to the colonial study of the tin industry in the fact that the industry was the largest during the colonial period in the Jos Plateau. There is also some justification for reviving the scholarship on the tin industry. Gonyok's study, like most dissertations, ends in 1960, leaving some fifty years of the post-independence period for updating. Freund's study extends to the 1970s, leaving a gap of about thirty years. Nankin Bagudu's edited work, although fairly recent, contains papers with little evidence of fieldwork. In another development, besides tin, several other minerals have been extensively mined. One of them is garnet. So far one scholarly paper has been written on garnet, which focuses on its history and general effects.[2] The following discussion on tin and garnet mining begins with the former.

THE TIN INDUSTRY DURING COLONIALISM

The British colonized the Jos Plateau between 1902 and the 1910s in the general belief that the area was rich in tin deposits. Before colonial rule the indigenous population used tin to paste metals, especially knives, against rust. Being a noncorrosive metal, the industrial world needed tin to produce cans, paste tubes, water pipes and jet aircraft engines. The Royal Niger Company started operations at Naraguta in 1903 and by 1914 there were 200 mining companies registered for tin mining activities on the Jos Plateau with a capital base of £10 million.[3] During the depression of the 1930s, however, the number of mining companies shrank from 139 in 1928 to 78 in 1931. This was owing to many of the companies amalgamating to survive. One product of these combinations was the Amalgamated Tin Mines of Nigeria (ATMN) formed in 1937.[4] During the heyday of colonial mining, mining locations included Dilimi, Rayfield, Bukuru, Kuru, Federe, Sabon Gidan Kanar, Gurum, Tenti, Bisichi, Barakin Ladi, Makafo and Gindin Akwati, as operations covered the then Jos Division. Extension centers were opened at Mangu, Manguna, Daffo and Bokkos among others. Between 1928 and 1938, mining leases acquired through Exclusive Prospecting Licenses (EPLs) accounted for 93% and 92% of mined land in the affected areas, mainly Jos Division.[5] By 1913 indigenous smelting furnaces closed down as part of the campaign against local entrepreneurs and producers. A paltry compensation of forty pounds per annum was paid to the smelters during their lifetimes.[6] By 1923, indigenous enterprises related to the tin industry were completely destroyed.[7]

European mining companies deployed high tech, heavy machinery onto the mining fields including the dragline, hydraulic monitors, jig plants, gravel pumps and one dredge installed at Dorowa.[8] Others included bulldozers, dumpers, scrappers, bucket wheel excavators, drills, pressure pumps and pipes.[9] Small-scale equipment was also used and included diggers, shovels and head pans for use by African laborers for surface and underground mining. The use of heavy machinery was incorporated in the 1930s as part of the companies' strategy to survive the depression.[10] The tin industry on the Jos Plateau became the largest employer of labor in colonial Nigeria and in Africa north of the Limpopo,[11] as shown in the following Table 2.1.

Table 2.1 African Labor and Tin and Columbite Output in the Jos Tin fields

Year	*# of African Laborers Employed*	*Output of Tin and Columbite in Tons*
1910	1,500	772
1911	5,832	1,809
1912	12,037	2,885
1913	16,883	5,531
1914	17,883	6,143
1915	14,316	6,910
1916	19,250	8,187
1917	21,817	8,314
1918	21,568	8,434
1919	22,976	8,169
1920	22,976	7,382
1921	14,982	7,239
1922	13,481	7,319
1923	19,124	8,372
1924	22,702	8,857
1925	24,727	8,937
1926	31,375	10,595
1927	37,115	11,509
1928	39,757	13,041
1929	38,678	15,335
1930	30,072	11,905
1931	20,763	9,800
1932	18,089	6,000
1933	20,710	4,956
1934	34,720	7,528
1935	40,053	9,044
1936	45,120	13,432
1937	49,509	14,873
1938	43,388	12,382
1939	44,072	13,434
1940	48,610	16,964
1941	57,827	17,040

(continued)

Table 2.1 (continued)

Year	*# of African Laborers Employed*	*Output of Tin and Columbite in Tons*
1942	71,511	17,972
1943	80,105	18,265
1944	70,289	19,313
1945	64,555	17,050
1946	55,138	15,802
1947	54,720	13,883
1948	52,893	13,836
1949	52,562	12,259
1950	51,962	12,232
1951	57,585	11,589
1952	60,651	12,711
1953	56,878	13,164
1954	50,246	13,010
1955	49,200	14,178
1956	50,432	14,670
1957	55,949	15,119
1958	34,427	13,951
1959	33,374	8,260
1960	33,819	10,011[1]

Sources: C. K. Gonyok, "A History of Labor in the Tin Mining Industry on the Jos Plateau" (PhD thesis) Table III, African Employees, Wages, Prices of Tin Metal Output, Value and Profits that accrued to British Tin Mining Industry, 1910–1938, 157; Table VII: African Employees, Wages, Tin and Columbite Output Value, Expenditure and Profits that Accrued to the British Tin Mining Industry, 1939–1960, 210.

Against the backdrop of the large land area covered by the EPLs, the use of heavy machinery, large labor force and failure to abide by extant mining laws, the environmental effects of tin mining were indeed negative and significant. One of the major negative impacts included damage to farmland. In the 1940s J. G. Davies performed a survey of the extent of damage to land owing to tin mining in Gyel and Zawan villages of present Jos South Local Government Council and concluded that 2,033.9 acres or 3.10% of the original arable land in Gyel had been destroyed permanently owing to a lack of land reclamation techniques. In Zawan a total of 10.8% of the land had been destroyed by mining activities. In Shen more land had been permanently damaged than in Gyel, as mining took up about 10%

of the land.[12] Gonyok has estimated that by 1939, 458 square miles representing 46% of Berom land had been alienated to miners. By 1955 about 590 square miles representing 54% of their land had been alienated.[13] Several other mined areas lack such surveys. For example destruction to land occurred, to varying degrees in Naraguta, Kuru, Heipang, Vwang, Foron, Gashish, Rop, Ganawuri, Kwal, Bassa and Jere districts among others,[14] but they have not been surveyed in the same manner as those areas covered by Davies.

In the 1940s new legislation required mining companies to reclaim mined lands through filling mining shafts, but this remained only on the books; in practice, the mining companies did not adhere to the laws. Instead, eucalyptus trees were planted on the few lands reclaimed in the mining areas; however, the trees grew tall and produced forests and inhibited agricultural practice, as no crops did well under their shade.[15]

POST-COLONIAL TIN MINING TO THE 1990S

Nigerian nationalism was political and not economic in outlook. Thus, independence did not lead to the immediate cessation of large-scale, expatriate tin mining activities on the Jos Plateau, and the former Jos Division in particular. Rather the tradition continued for a considerable time thereafter. Major companies continued in this respect including ATMN, Exlands, Jantar, Kaduna Prospecting Company and Gold and Base. Between 1960 and 1984 the major companies were in full operation although the quantity of tin mined since 1970 declined at a regular rate. The 1960s were good for the tin industry. Alhaji Aliyu, Makaman Bida, the honorable Minister of Finance in the Northern Regional government in 1962 stated that

> The mining sector of the economy, the improved conditions I was able to report last year have been maintained. All restrictions on the output of tin having gone, its output has been increasing, and the world price for tin reached higher points this last year than for some years output. On the other hand market for columbite has been dull.[16]

Confirming this position, Freund has shown that in the 1960s, tin production increased and mining companies resorted to the use of the draglines and other heavy machinery that were turned off in the 1950s in the face of the upsurge of private miners. At the same time the number of tributers[17] plummeted from 20,677 in 1950–51 to 2,797 ten years later in 1960–61 owing to pressure from the mines department that their activities were tantamount to gross exploitation of labor and were destructive to the mines. Conversely, within the category of hand laborers the number of contract workers grew.[18] The independence period also witnessed better opportunities for indigenous (Nigerian) miners. By the eve of independence in 1959, eleven Nigerians were licensed to mine tin but they were responsible for only a small quantity of tin obtained via contract mining. In the same year, another development occurred in favor of indigenous miners. As a sort of encouragement to them, the government opened a dredging plant, the effect of which was to free small-scale operators from dependence on company facilities.[19] After the termination of civil rule, the mining industry increasingly came within the purview of public enterprises. In 1967 the Northern Nigerian Development Company acquired 16 percent of the shares of Makeri Smelting Company. In 1972, the Government established the Nigerian Mining Corporation (NMC) with the mandate of investigating the possibilities of underground mining. In 1974 Gold and Base acquired 40 percent of the shares of NMC and planned to enter a joint venture with the NMC in underground mining. The Nigerian Mining Corporation acquired 40 percent of the shares of ATMN in 1976. The Amended indigenization decree of 1977 gave 60 percent of the shares of ATMN to the government. Loans dispersed to indigenous miners intended to improve their lots in the mining business, but given the indigenization law (the Nigerian Enterprises Promotion Decree of 1972) and the insufficiency of the loans, indigenous miners were unable to compete with the foreign companies.[20]

Post-colonial mining depended on a variety of different technologies and the technology used in the colonial period continued in use until the 1990s. Thus, different companies continued to use heavy mining equipment like the dragline, dredge, hydraulic monitors, jig plants, gravel pumps, bulldozers, dumpers, scrappers, bucket wheel excavators, drills, pressure pumps and pipes up to the 1990s. ATMN alone used several draglines of different sizes numbered one to seven. They worked various locations including Sabon Gidan Kanar, Bukuru, Barakin Ladi; Ganan Daji, Kubuwa, Ngel, Foron and Major Porter (on Bokkos Road) among others.[21] The dragline is a massive machine of about 100 meters in length. The control cabin is about the size of a very large hall. Alhaji Adamu claims it has twenty-four gears and two control panels to the left and right of the operator. The right control panel is for digging, whereas the left panel is for rotating the digging compartment in different directions. The dragline digs deep down for top and bottom washing. Top washing reaches about 75–150 feet at which the first level of tin is found. The tin at this level is lower quality. Bottom washing reaches between an additional eight feet at which the water in the pit changes color; clean water being reached. At this level very high quality tin, tantalite and columbite are obtained. The dragline operates day and night.[22] According to Toma Jang Davou, "a dragline could shift 200 cubic yards per hour or 2000 cubic yards in ten hours and was equivalent to a gang of 1,330 men. The dragline could make over work [sic] sixteen or if necessary twenty-four hours as a day [*sic*] while labourers could only carry on head panning working during the daylight hours."[23]

The dredge was a large piece of equipment (larger than the dragline) that floated on water, operating only along rivers and was used at Dorowa valley. It dug and processed different minerals according to their types on its own; it mined fill concentrates, i.e., tin, columbite, ilmenite, zircon and monazite.[24] During the 1970s the tin industry used a mixed grill. Heavy machinery was becoming old and less productive; but tin prices rose. It was also the period of the oil boom, which eventually led to a decline in the industry.[25] However, the tin industry remained the largest productive establishment on the Jos Plateau for a long time after independence. The scale of operations in the tin industry after independence can be deduced from the size of African labor employed and the output of tin, respectively, as shown in Tables 2.2 and 2.3 below.

Although there are indications that large-scale technology continued to be used up to the 1990s,[26] the use of low/simple technology began to dominate the mining industry since the late 1980s as large-scale mining machines became old and output collapsed. Sam Colom argues that most of the imported heavy mining equipment became old and less productive by the mid-1980s as the expatriate firms failed to import new equipment or fix old ones. In his words, "Europeans calculated the useful life span of their

Table 2.2 Number of African Workers on Minefields

Year	*No. of African Laborers Employed*
1955/56	51,201
1956/57	50,888
1957/58	46,211
1958/59	28,342
1959/60	28,928
1960/61	34,543
1961/62	39,101
1962/63	42,639
1963/64	44,440
1964/65	47,799
1965/66	55,290
1966/67	58,417
1967/68	55,586
1968/69	48,586
1969/70	50,675
1970/71	52,188
1971/72	52,639

Source: Bill Freund, *Capital and Labor,* Table 8.6, Number of African Workers on the Minesfields, 210.

Table 2.3 Nigerian Cassiterite Production 1963–1985

Year	*Tons*
1963	12,250
1964	11,800
1965	12,900
1966	13,950
1967	12,500
1968	13,450
1969	12,000
1970	11,850
1971	9,750
1972	9,500
1973	8,200
1974	7,400
1975	6,300
1976	5,300
1977	4,500
1978	4,200
1979	3,700
1980	3,750
1981	3,250
1982	2,400
1983	1,950
1984	1,850
1985	1,400

Source: Engr. Sam G. Colom, "Solid Minerals Development in Plateau State, The Journey So far" (paper presented at the Raw Materials Research and Development Council in Collaboration with the Plateau Minerals and Development Company Seminar on "Exploitation and Utilization of Mineral Raw Material Endowment of Plateau State for Cottage Level Development," held at the Tudun Wada Government Lodge, Jos, August 17th, 2006.

equipment and made sure they (machines) worked for that life time and did not import new equipment or parts," having become less interested in the tin industry owing to the indigenization law of the 1970s, which made them minority shareholders. Thus, to him, the indigenization law killed the mining industry.[27] The indigenization decree was promulgated by the military government of General Yakubu Gowon in 1972 to improve Nigerians' participation in the Nigerian economy. It classified businesses in Nigeria into Schedules I, II and others. Businesses in Schedule I, which were low-scale, less capital intensive, were reserved 100% for Nigerians; those in Schedule II, which were medium scale, were to allocate 40 percent to Nigerians. The categories called others were large-scale, high technology and were not affected. It was reviewed in 1973 and 1976. The revisions removed the category others and classified all business by creating Schedule III, which was to reserve 40 percent to Nigerians, as Schedules I and II were expanded with Schedule II forced to allocated 60 percent of the shares to Nigerians who must retain 100% shares in Schedule I.[28] But it was not only the indigenization law that destroyed large-scale tin mining. To this factor should be added the collapse of tin output. The decline started in the late 1960s and the industry never recovered. Hence from a capitalist point of view, investing in heavy machinery in the period of decline did not make economic sense. Indeed between 1985 and 1998 tin output, as represented in a chart provided by a mining firm, fell to less than 1,000 tons per annum.[29]

Although Consolidated Tin Mines (CTM) succeeded the expatriate firms (ATMN, Bisichi-Jantar Limited, Gold and Base Metal Mines of Nigeria Limited, Ex-Lands Nigeria Limited and Kaduna Prospectors Nigeria Limited) being incorporated in 1988 as a public company,[30] no new equipment was imported. As years passed, the draglines were sold off except the last one at Barakin Ladi, which stopped working more than ten years ago.[31]

Thus, the following simple tools that were used as subsidiary to the heavy machines in the colonial period dominated the post-colonial tin industry as the major mining tools. They included a bucket wheel excavator, diggers, shovels, wheel barrows, head pans, calabashes, spoon and water pump/generator (small firm structure).[32] The bucket wheel excavator remains the single largest equipment on the mines. CTM is currently using it for its skeletal operations at Rim, as private miners use it in their dry season operations used mainly in the *loto* system in which a well is dug to a certain depth—60–100 feet using diggers and shovels—whereas a bucket is used for evacuating the earth from the *loto*. At the required depth where tin is found, digging is done horizontally. The *loto* system is done in the dry season mainly (after the New Year festivities) because water levels would have deepened. In this technology about 10 persons are involved as six to seven miners could be working underground digging and loading the earth into a bucket while those at the surface wheel it out and pack it. Subsequently, the miners might hire a truck to evacuate the earth to a washing place. When a *loto* is known to contain tin, parallel groups could also set up their bucket wheel equipment and engage in mining by creating their own *loto*.[33]

Abubakar Ya'u described the functions of each of the other equipment as follows. A digger is a two edged metal tool fixed on a wooden handle. Diggers are used to dig the earth and remove the top soil in search of tin. A shovel is used to remove the dug earth into a water channel or as a container.

Head pans were used in the colonial and early independence periods to pack dug earth to create artificial mounds and tin to washing sites. They have now fallen into disuse. Calabashes are used to wash tin by separating it from earth and stones. Tin is heavier than earth and usually settles at the bottom of the calabash while the earth floats on top and is poured away. Table spoons are used to scoop identified tin by river beds.[34] Water pump generators were introduced in the last two decades. Dry season farmers first used them, and then also tin miners. Medium scale miners use water pumps in the post-independence period to pump water to a mining point. A channel is created for the water sometimes over a distance of 100 meters. The water is used for two purposes, first to expose tin from lumps of earth poured into a specially prepared channel hedged on both sides with grass or corn stock. The second use of the water is for washing the tin packed from the channel into the washing place. In this type of mining activity three to five persons could be involved, as some entrepreneurs hire laborers to work for them. In this case the land (former mining mound) is hired from the land owner who is paid per day, as is the case with hired laborers.[35] The individual miner uses digger, shovel, bucket or wheel barrow (to move the earth to the washing site) and calabash.[36] Such a person operates by a river where the mine is very close or by the riverside where the washing of the tin takes place. Most mining activity is essentially based on open cast or surface mining. This is different from underground mining, the *loto* type. In the colonial period the *loto* type of mining was called *Gudun lamba* in Hausa, as it was done by illegal miners or tin thieves who did so to steal tin from an active site.

Much of the mining taking place since the 1990s is located on old mining sites and riverbeds. Miners on old sites dig gangue[37] left by heavy machines because they know that tin was obtained from the soil and that it is possible to still find particles of the product in the gangue. A majority of the sites I visited and confirmed by informants at Barakin Ladi, Rayfield, Sabon Gidan Kanar use this tactic, as most of the young miners learned the art from their fathers or older persons with mining experiences.[38] In some cases the gangue is hired by the miners for some specific fee. At Barakin Ladi, there is a gangue at a place called Kuwait, which is worked daily by about four miners. Davou Pam who claims to be the entrepreneur for whom three boys work asserts that he pays the sum of N500 as a hiring fee daily on the gangue, whereas he pays his three laborers N1,000 daily. Where fresh land is involved, Abubakar Yau claims that between N50,000–N200,000 could be paid as compensation to the land owner. Nothing is paid when it is family land mined by the land owner or members of his family.[39] In dry season, non gangue mining, the miner establishes the presence of tin in the area by collecting samples of the soil and washing it.[40] Mining in the rainy season is not necessarily on an old mining site but by riverbeds or on farm land where rain water has exposed the mineral on the top soil.

The majority of miners who use simple technology are farmers and a few are students. This category of miners takes to mining only as a part-time—dry season or leisure time—occupation. They win very small quantities of tin and columbite sometimes daily but most often after a day or two and sell it to local agents or accumulate up to three to five salt bags of about ten kilograms each during the week and take it to Bukuru or Jos to process and sell. None of my informants knew the uses of tin. They only knew the local markets. They use money from selling tin or columbite to buy chemical fertilizers, and other life necessities such as clothing, drugs, and also to buy drinks. Students use their money to supplement the cost of their education. The name of one processor and buyer was on the lips of many of the miners. Chike is the name. His processing plant is at Bukuru. It was only in Bisichi that I met someone like Salisu Adamu who claims mining is his only profession. There are many people, according to him, who are also professionals. This category of miners wins a large quantity of tin and columbite. I saw over ten bags of ten kilograms of unprocessed tin and columbite mined by Mr. Adamu alone on the day of the interview. Mr. Adamu showed me his car, which he has bought from the proceeds of mining.[41] He claimed that four other successful professionals also had cars and had built their own residential houses. However, as we shall argue later, whereas the professionals gain, the community loses in terms of easy access to water supply.

GARNET MINING

Between the 1980s and 1990s garnet was extensively mined in the Jiblik communities of Gibang and Guza, Lankan District of Pankshin local government area (LGA) of Plateau State. The former Pankshin LGA (now Pankshin and Kanke) is endowed with various solid minerals including among others aquamarine and tourmaline (in Kabwir); garnet (in Jiblik); topaz (in Dawaki-Kabwir); tantalite (in Pankshin and Fier); quartz (in all the districts); glass sands (in Pankshin); granite (in Pankshin); charnockite (in Dungum-Ampang East); and mica, tin and columbite (in various districts).

The beginning of garnet mining in Jiblik is associated with Mrs. Nanlep Birbu who in the mid 1980s made her brother, Boniface Birbu, her agent in the village having come across the garnet business either in Jos, Zaria or Kaduna as she traveled to these cities. Boniface Ropshak made Mr. Bala and Mr. Benle sub-agents in Guza and Gibang areas, which are their respective domiciles.[42] Villagers identified garnet from other stones because of its dark brown or black color and the fact that it glitters. At this initial stage tiny pieces of garnet were picked from the ground, with underground mining yet to be introduced. Unlike tin, which has a pre-colonial history of production and use, garnet mining was a post-colonial venture and the villagers had no idea what exactly was its use. Those who cared to ask were told that it was used to manufacture jewelry items. Garnet was produced for the external market from the start. Eventually, miners discovered a market for the product in Jos and surface mining spread in Jiblik village by attracting miners from various villages within and outside Jiblik and Pankshin LGA and also by attracting the entry of mining companies.

Two companies were said to have been involved with garnet mining in Jiblik. These were A. M. Kalambu and Company Ltd. and Dankoli Mining Company, belonging to Alhaji Hashimu Dankoli.[43] Of the two, A. M. Kalambu had a more serious presence and impact, whereas Dankoli was hardly mentioned by the villagers. In 1987 A. M. Kalambu and Company Limited of Dilimi, Jos, applied for and obtained prospecting rights over the garnet-bearing areas of Gibang. The company was granted a prospecting license and "C" class mining lease number 19834. The license and the lease covered garnet, ruby and sapphire and lasted for a period of ten years effective August 28th, 1988. The land area covered by the lease was 183.46 acres at Gibang.[44] The then ministry of Petroleum and Mineral Resources approved the license and lease on September 15th, 1988. Subsequently, the company took measurements of the land covered by the lease and paid out compensation to affected peasants. Table 2.4 below shows the names and amounts of compensation paid to land owners in 1988 and 1992.

Against the background of the then existing policy of Structural Adjustment Program (SAP), the amounts paid as compensation to land owners were very low. During the course of carrying out this research, I made several visits to the company office at Dilimi but failed to get any official to give me useful information on the rationale behind the figures or to confirm or deny them. I was also unable to secure a meeting with Huseini, the company's agent, as he did not reside in any part of Mupunland.[45]

Table 2.4 Compensation Paid to Land in 1988 and 1992

S/No.	*Name of Payee*	*Amount (1988) (N)*	*Amount (1992) (N)*
1	Tohomdet Dachal	1, 009.00	405.56
2	Danladi Abalis	567.50	—
3	Joshua Dahar	569.00	—
4	Kumlap Rumji	—	647.10
5	Boniface Wukden	160.50	—
6	Biyangshak Dashwep	28.50	788.70
7	Pelden Dasen	—	488.17
8	Yakubu Dashem	—	559.08
9	Daniel Dawuk	—	112.85
Subtotal		2,333.10	3,002.46
Total			5,335.56

Source: Mupun Cultural and Development Association (MUCDA) Joint Action Committee Report, "Report on mining activities in Mupunland, November 1993–March 1994," 3–5.

N = Naira

The entry of A. M. Kalambu Ltd. led to a number of results: first, Mrs. Birbu's monopoly ended. Second, the company purchased and distributed shovels and diggers to laborers and imported laborers from former Jos Division and Hausaland with experiences in underground mining. Consequently, underground mining went on simultaneously with surface mining pulling a large labor force comprising people from Jiblik and nearby villages and those from other places in Plateau State and other areas. Underground mining called *loto* went to a depth of between four feet to thirty feet at which the highest grade of garnet called *Jambo* was obtained. The ethnic backgrounds of imported and independent laborers included Mupun, Berom, Hausa and Kanuri, Igbo, Yoruba, Tiv, Idoma, Ngas, Mwaghavul, Miship, Pyem, Afizere and Fulani from other parts of the Jos Plateau and Nigeria in general. Non-Nigerians included Cameroonians, Chadians, and Ghanaians. Laborers imported by the company were obliged to sell all garnet they mined directly to the company. Independent laborers were not under such obligations. Third, the company's principal agent, Huseini, appointed sub-agents from the Mupun ethnic group to buy garnet from independent miners and company laborers.

According to Mr. Monday Josiah the use of heavy mining machines was contemplated but that there was opposition from the communities and the chiefs who believed that heavy equipment would lead to the displacement of villagers and would be more destructive to the land. There is some justification for the communities' rejection of the use of heavy machinery. This is the fact that they are basically agrarian, using small peasant holdings inherited from their forebears, as the rocky and hilly environment was an additional constraint to land use. Consequently, small equipment including hoes, metal hoe heads, small hoes (especially the thick type called *magirbi* in Hausa), axes, spears, knives, diggers, shovels, calabashes and fertilizer bags among others became the major tools.

Company miners and especially independent miners did not work exclusively in areas covered by leases to companies, as land owners allowed independent miners to mine their farms according to an agreement between the miners and the landowners.[46] Thus, apart from Gibang and Guza, mining also took place in Dangper, Koyillum, Tuwat, Togchiot, Amshang, Koplong, Ninik, Jinin and Juruk in various parts of Jiblik. During my field trip to the various mining sites at Gibang and Guza in 1996, I saw up to 200 miners on one location alone. The implication is that when all the numerous sites are taken together the average labor force per day would be slightly over 1,000.

Garnet mining declined in the closing years of the 1990s and by the beginning of the twenty-first century garnet mining had virtually come to a close, as the companies ceased operations leaving only individuals to engage in skeletal activities. According to the villagers, the mineral had been exhausted "*nghik ni ki wa*" (literally the mineral has returned home). Besides, some of the leases had expired. A. M. Kalambu's lease expired in

1997, whereas Dankoli's was subject to a three-year renewal with the communities based on the initial operational agreements. Furthermore, tension between the villagers and miners was becoming too frequent owing to arrests and detentions, attacks on illegal miners by aggrieved landowners and the persistent opposition to the mining activity by MUCDA from the mid 1990s. This will be further elaborated in the section dealing with reactions to mining activities by affected communities.

ENVIRONMENTAL IMPACT OF POST-COLONIAL MINING

Research shows that there is a strong relationship between the technology used and the extent of negative environmental effects. The larger in scale the technology, the larger the quantity of land affected; the obverse is the case with small-scale, low technology. The 1930s to the 1980s was a period of large-scale and heavy technology and was responsible for much of the environmental damage in the tin-bearing areas of the Jos Plateau.

The greatest environmental effect of tin mining using heavy machines was the destruction of vast quantities of land. Emmanuel J. Dasat, a former Director in the Plateau State Ministry of Environment and Mineral Development, claimed that by 1983 when most of the large mining companies folded up operations, mining operations had "left in their trails not less than 316km^2 of very unproductive land, 35 active gullies of over 7km in length, over 1000 un-reclaimed mine ponds and an almost unaccountable number of dry depressions, pits, and scars."[47] D. D. Dabi has argued that the estimate of 316 km^2 is an old one and that the new estimate is 325 km^2 concentrated in the Jos-Bukuru-Barakin Ladi-Ropp axis.[48] One aspect of the destruction of the land was that mined areas lost their nutrients. In 1949, the Mine Lands Reclamation Unit (MLRU) was established to reclaim areas not covered by the Mining Act of 1946. The hope was that the reclaimed area would be used for agriculture after filling the pits and ponds. However, "The inherent low fertility of the spoil and consequent need for input (at that time) of totally uneconomic levels of both inorganic fertilizers and manures caused the MLRU to abandon reclamation for agricultural purposes and to instigate the policy of planting eucalyptus in the Mine Reclamation Areas (MRAs)."[49] Indeed, "The results of soil analysis for both reclaimed and unreclaimed spoil . . . emphasize low nutrient status of the spoil materials which caused the MLRU to abandon its policy of reclamation for agricultural purposes. For example, the nitrogen and available phosphorous levels are so low that crop growth would be seriously affected. Furthermore with pH levels below 5, generally nutrient availability would be restricted to the point that both cereals and vegetable yields would be greatly reduced."[50]

The devastation of large quantities of land created land shortage. According to Alexander, the evidence of land shortage was that villagers were moving into MRAs for their farming activities.[51] Another effect of large-scale mining is environmental pollution. According to John Dung Gwom, other negative effects of tin mining include radioactive emissions capable of causing brain damage and mental deficiency and water, air and soil pollution from base metals released through mining.[52]

Another effect is alteration of the landscape. Mangvwat suggests that over half of the landscape of the former Jos Division was destroyed, as the vegetation and soils of the area were harmed. In his words, "The sinking of large, deep and open shafts and the removal therefrom of millions of tons and volumes of earth, loaded onto the surface in mounds, created a totally new landscape in the numerous areas affected. All over the mining districts, artificial lakes, ponds, gorges and mounds were created over original farmlands. Dams were also constructed . . . for the purpose of generating electricity which the tin industry needed."[53] Besides, there was massive felling of trees for energy and construction purposes, thereby rendering the soil vulnerable to erosion, as erosion caused by gullies from artificial mounds created by tin mining activities did occur on a large scale.[54]

Although the colonial period has been much studied, there has been no data documenting the specific quantity of land damaged at that time; thus, all the above impacts are for the colonial and post-colonial periods put together. Several actions have been taken and useful suggestions given on the damage done by mining using heavy machinery. For example, in the colonial period, the government was forced, in 1946, to draft legislation requiring reclamation of mined lands. Indeed, prior to that time, "Restoration conditions were not mandatory in lease titles until 1946."[55] The Restoration Clause was to be effected on lease areas declared as congested areas. By this provision, the Northern Regional government was responsible for reclaiming mined lands, whereas the mining companies were to restore subsequent mined lands leased to them. In 1954 the regional government earmarked the sum of two million pounds to reclaim Jos Division and parts of lands adjoining Bauchi province declared to be congested areas. A Mines Land Reclamation Unit (MLRU) was established at Bukuru and claimed in a gazette to have reclaimed 60 mine areas totaling 3,054 acres.[56] The MLRU became moribund with the civil war but was revived by the Benue Plateau State with a four-year allocation in 1971–74. The reclamation burden soon became too heavy for the state government, which sought the cooperation of the Federal government that resulted in the setting up of the Joint Consultative Committee (JCC), with the Assistant Director of Mines, Jos, as chairman. The MLRU under the JCC received some federal support leading to the completion and handing over of the Barakin Ladi Reclamation Project (awarded to ATMN) and execution of the "Mystery Dam" project at St. Murumba College (awarded to Roads Nigeria). Upon restoration of democracy in 1999, the Ministry of Environment and Mineral Development was created in August with the mandate of, among others, "land reclamation, erosion and flood control."[57] Despite these efforts, the problems of environmental damage still persist owing to unsuccessful reclamation attempts. The challenges of gangue, ponds, lakes and gorges left by large-scale mining activity have therefore attracted suggestions including reclamation of areas where the ponds pose a danger to life and the use of the 1,000 ponds for all year round irrigation for agriculture, especially given the shortage of employment opportunities, so the government should encourage small farms and farmers. Other ponds such as the one in Rayfield are used as recreational facilities.[58] Some of these suggestions are already being implemented. Another evidence of the implementation of some of these suggestions is that the Rayfield Resort has been established as a recreational facility, whereas the pond at Bukuru is being closed. Villagers take advantage of the lakes as sources of water for dry season farming. When I asked them their opinion on the closure of the lakes, they said they would prefer them left unclosed so that their livelihoods can continue to be guaranteed. Sam Colom partially supports this view arguing that closing all the lakes and ponds would incur unnecessary cost, but he contends that those within towns should be closed. His argument makes sense because Plateau state is an agrarian economy and products from dry season farming stabilize the food economy. Moreover, as argued by the M. J. Alexander, mining destroyed the fertility of the soil and reclaiming it for agriculture requires re-fertilization.

For the period after the 1980s, when light and local equipment have been in use and old mining sites reworked, the environmental impact is insignificant but not zero. Land degradation has continued, as gangue, fresh farmlands, rivers and water courses are occasionally devastated. Mining using the bucket wheel excavator continues to leave pits that pose as danger human and animals, as people and animals can fall inside. At Bisichi the stream where ATMN used to flush out tin waste is permanently mined and on a daily basis, as professionals and part timers alike congregate on the stream. According to Salisu Adamu up to 20 miners work the stream daily. I saw numerous washing places, which Salisu said belonged to different individuals. The stream that is permanently mined is no longer used as a source of water supply. In Barakin Ladi a gangue and a gorge are worked permanently, thereby expanding the stream and denuding the land.

In the garnet areas, mining led to temporary economic boom and social upliftment in the lives of agents and sub-agents and successful miners. On the other hand, the physical environment was negatively affected; 183.46 acres were granted as mining leases to mining companies, but the company of A. M. Kalambu failed to confine its operations to its lease. Besides, landowners also granted concessions to independent miners. The effect of these developments was the spread of mining over a large expanse of fields. Naturally the soil was disturbed through scratching the surface and digging including underground mining. There was also the felling of trees, and rolling off of stones, all of which left the soil with trenches, pits, holes and bare soil, which became vulnerable to erosion. Grazing and hunting grounds were also destroyed, as there is no waste land in Jiblik. The mount (*Zuul)* Gibang was defaced; thus its scenic beauty and the tourist potential were destroyed. Kalambu's company got granted on its lease issued by the Federal Ministry of Mines and Power a "no-restoration" (of the soil) clause,[59] contrary to the mining laws of Nigeria. The mining laws of Nigeria stipulate that the ground surface of a mining area must be protected, fenced and filled up whether permanently or temporarily abandoned to avoid the danger of people falling into the pits or trenches accidentally.

Rivers and streams were also intensively mined, which caused them to expand unduly and submerged farmlands by their banks. The expansion of the watercourses caused an additional transportation problem. The rivers and streams became more difficult to cross during the rainy season, as much effort put in by the communities to make them motorable became wasted owing to roads being washed. An additional problem was water pollution occasioned by the mining of water courses and their use for washing Garnet, which is radioactive. Furthermore the mining of water courses violated the mining law of Nigeria, which provides that "No person shall in the course of mining or prospecting operations or in any works connected therewith pollute or permit to become polluted the water of any watercourse."[60]

REACTIONS TO MINING POLLUTION

Colonial tin mining elicited vibrant, sometimes violent reactions from different segments of society. On one hand there were series of wage and forced labor related protests by miners.[61] On the other hand the negative environmental effects of tin mining using heavy machines drew the attention and reaction of the Berom, who were the most affected. Their reaction was mass and elite based. On the side of the peasants, reactions sometimes took the form of delays in paying tax and attacks on tin miners as happened in Gyel in 1942, which forced the government to direct J. G. Davies to assess the extent of land damaged and for the purpose of land compensation. Further reactions came from the elite-led Birom Progressive Union (BPU) formed in 1946, as it took up the issue of land compensation along with political demands including representation in Native Administration. The demand for land compensation was met in 1948, but struggle over control of land continued and in 1951, the BPU led peasants to disrupt mining activities on one site. The BPU also successfully fought against schemes to resettle Berom people who lost their lands at Sabon Zawan near Jema'a in the 1950s.[62]

Whereas not all the mined lands were reclaimed, the reclamation project was abandoned because of the low nutrient status of reclaimed land. There has been no violent reaction from affected communities in post-colonial times for a number of reasons. One of the reasons is the fact that compensation of some sort, sometimes ridiculous, was paid to affected land owners by the companies.[63] Another informant attributes this to the rather peaceful disposition of the people.[64] Other reasons are alluded to by the ATMN, one of the largest companies that operated in the then Jos Division. According to the company

> Often the after effects of the mining industry are held against the mining companies, because it is said that farmland has been spoilt by mining, but it must also be remembered that many of the mining sites were not originally used as farmland e.g. the Sabon Gida valley was a huge swamp. Restoration conditions were not mandatory in lease titles until 1946.
>
> In addition some of the best market gardens are in the areas that had been mined and subsequently restored. There are also the eucalyptus plantations on old mining areas, which yield valuable harvests of timber.[65]

The factors alluded to are first, the fact that some of the sites were not active farmlands at the time of their being leased out. Second, the reclamation of some mined lands seemed to have given landowners the hope that one day all lands affected would be reclaimed. After independence another type of reaction was in the form of agitations for reparations, ecological funds and reclamation/rehabilitation of mined lands led by the elite.[66] Some ecological funds have been released but indications are that such funds have been mismanaged in the last decade.[67]

In the garnet mining areas, reactions were both violent and civil. Indeed, garnet mining generated crises in the village similar but not to the same extent as the ones in the Niger-Delta. The vast majority of Jiblik people were sad at the wanton destruction of the village farmlands because land does not belong to individuals as private property. The present holders are but trustees on behalf of present and future generations. The people were also unhappy with the operations of A. M. Kalambu and company, which tried to prevent open free competition. Unfortunately, like in the case of the oil producing communities of Niger-Delta, the youth and the chiefs did not support a common approach to the problem. For example, whereas the youth preferred to take radical steps even if illegal to end the menace, the chiefs preferred a gradual, diplomatic approach, which the youth interpreted as a sign of weakness and complicity with the company and independent miners. Tension was heightened between the company and the entire Mupun community when one villager, Mr. Gwol from Abwor/Dyis village, was killed in a police cell on allegations of garnet theft. In 1995 it took the youth and the chiefs several meetings to resolve their differences. Then the differences reemerged as clashes frequently occurred between the company and unauthorized miners on one hand and farm owners on the other. Most landowners actually objected to the mining of their lands and spent a good deal of time patrolling their lands to stop illegal mining. Some of the miners were so persistent that on numerous occasions landowners chased them out by pelting them with pebbles and sticks. Such unauthorized miners committed acts of trespass in the daytime but most often in the night or during periods of suspension of mining (the farming season). Some of the miners had prior permission from landowners but exceeded the area covered by the agreement, which led to conflicts. Most often the clashes and conflicts led to police action on the initiative of the miners, whom the landowners accused of having the backing of the police.[68]

NOTES

1. See among others, Penelope Bower, "The Mining Industry," in *Mining, Commerce and Finance in Nigeria,* ed. Margery Perham (London: Frank Cass, 1948); Bill Freund, *Capital and Labour in the Nigerian Tin Mines* (London: Longman, 1981); C. K. Gonyok, "A History of Labour in the Tin Mining Industry on the Jos Plateau" (PhD thesis, History Department, ABU, Zaria, Nigeria, 1986); J. J. Grace, "Tin mining on the Jos Plateau before 1920," in *Studies in the History of Plateau State, Nigeria,* ed. Elizabeth Isichei (London: MacMillan, 1982); Nankin Bagudu, ed., *Land Desecration and Environmental Degradation on Nigeria's Jos Plateau* (Jos: League for Human Rights, 2001). There are also numerous Masters and BA Dissertations on the subject in the University of Jos, creating a historiography of what Monday Mangvwat has called "tin determinism." See Monday Yakiban Mangvwat, "Tin Mining and Environmental Degradation on the Jos Plateau, 1902–1994," in *Land Desecration and Environmental Degradation on Nigeria's Jos Plateau,* ed. Bagudu, 61.
2. S. U. Fwatshak, "Garnet Mining and Its Impact on Jiblik Communities From the 1980s to the 1990s" in *Historical Research and Methodology in Africa,* ed. Mike O. Odey, John G. Nengel, and Okpeh O. Okpeh (Makurdi, Nigeria: Aboki Publishers, 2007), 270–292.
3. Leonard Plotnicov, *Strangers To the City: Urban Man in Jos* (Pittsburgh: University of Pittsburg Press, 1967), 34; Bill Freund, *Capital and Labour,* 38, 111.
4. C. K. Gonyok, "A History of Labour," 138, 153.
5. Bower, "The Mining Industry," 7–8, 14; S. U. Fwatshak, "A History of Private Indigenous Entrepreneurship in Jos Metropolis, 1902–1985" (PhD thesis, University of Jos, 2002), 153.
6. SNP/17/3/29195 Tin Smelting in Lirue-n-Dalma by Natives; Prohibition of.
7. Bower, "The Mining Industry," 4.
8. Amalgamated Tin Mines of Nigeria Limited, "The Plateau Minesfield: What you should know about Nigerian Tin Production" (n.d.), .6; Toma Jang Davou, "Tin Mining and the Ecology of Plateau State" in *Land Desecration and Environmental Degradation on Nigeria's Jos Plateau,* ed. Bagudu, 32; Gonyok, "A History of Labour," 147, 153.
9. Sam G. Colom, interviewed in his office at CTM, Bukuru, 01/09/09; Amalgamated Tin Mines of Nigeria Limited, "The Plateau Minesfield" (n.d.), 6.
10. Gonyok, "A History of Labour," 153.
11. W. M. Freund, "Labour Migration to the Northern Nigeria Tin Mines, 1903–1945," *Journal of African History* 22, no. 1 (1981): 78.
12. Davou, "Tin Mining," 34–35.
13. Gonyok, "A History of Labour," 191–192.
14. Mangvwat, "Tin Mining and Environmental Degradation," 63.

15. Ibid., 64–65.
16. Alhaji the Hon. Aliyu, C. M. G., C. B. E., Makama of Bida, *Northern Nigeria Legislative Parliamentary Debates* (Hansard), Saturday, March 17, 1962, 43.
17. Tributers were indigenous tin miners of African descent who were permitted by the expatriate mining companies to win and sell tin to them.
18. Freund, *Capital and Labour,* 209.
19. Ibid., 217.
20. Ibid., 224, 223.
21. Alhaji Adamu Kano, 61 years, interviewed at his house Mining Quarters at Barakin Ladi, 01/2/09. He started work as a messenger and then became garden boy and finally dragline operator. He has operated ATMN's 7 draglines in the different locations at different times before his retirement; Sam G. Colom, interviewed January 9, 2009.
22. Alhaji Adamu states on some occasions while operating the dragline at night sometimes he saw spirits in various forms on the spot. One was in the form of a woman carrying a baby on her back, the other was in the form of ducks, whereas the third was in the form of a black dog. What he did in these circumstances was to shift the digging compartment away from the figure and temporarily stop working until the "spirit" had passed by or left the spot.
23. Davou, "Tin Mining and the Ecology of Plateau State," 32.
24. Sam G. Colom, interviewed January 9, 2009; Alhaji Adamu Kano, interviewed 01/02/09.
25. Freund, *Capital and Labor,* 224–225.
26. Salisu Adamu interviewed at Bisichi 01/12/09 claims the gravel pump at Bisichi stopped working in 1999; whereas Samaila Alhassan interviewed at Sabon Gidan Kanar, 01/10/2009, claims that the last CTM dragline line in Sabon Gidan Kanar stopped working about 10 years ago.
27. Sam G. Colom, interviewed January 9, 2009.
28. For a full discussion of the indigenization law, see Thomas J. Biersteker, *Multinationals, the State and the Nigerian Economy* (Princeton, N.J.: Princeton University Press, 1987).
29. "Nigerian Tin ore Production–(1963–1998)," provided by Engineer Sam G. Colom during an interview session I had with him on January 9, 2009, in his office (CTM Headquarters, formerly ATMN headquarters) at Bukuru.
30. This information is contained in a leaflet entitled "Consolidated Tin Mines Ltd, Keffi Yard-Bukuru, Jos Nigeria" (n.d.).
31. Abubakar Yau, 33, interviewed on his farm (12/31/08; Ezra Pam, 23, interviewed on site (washing place), 12/31/08; Gyang Pam, 26, primary school leaver; Musa Nura, 15, primary school leaver; Davou Pam, 23, primary school leaver; Davou Pam, 30, secondary school leaver, all interviewed on site (digging) 12/31/2008; Adamu Kano, interviewed at Barakin Ladi, 1/2/09; Sam G. Colom, interviewed January 9, 2009; Samaila Alhassan interviewed at Sabon Gidan Kanar, 01/10/09.
32. Interviews with Abubakar Yau, Ezra Pam, Gyang Pam, Musa Nura, Davou Pam, and Davou Pam; Chuwang Dung, 23, primary school leaver, interviewed 01/10/09; Idowu Fedelis, 22 (Plateau Polytechnic Student) interviewed on site at Sabon Gidan Kanar, 01/10/09; Emmanuel Boyi, 30, Dung Choji, 30, both interviewed on site at Barakin Ladi 1/12/09; Salisu Adamu, 35, Ibrahim Gwom, 28, Terry Ephraim (student of Mineral Resources, Plateau Polytechnic, Barakin Ladi). The three were interviewed at Bisichi 01/12/09. The first two were on site, the third I gave a lift and we got discussing.
33. Emmanuel Boyi ,30; Dung Choji, 30, interviewed at "Kuwait," Barakin Ladi, 1/11/09. They both have experiences in *loto* mining according to their testimonies.

34. Abubakar Ya'u, interviewed 12/31/08.
35. Davou Pam, 30 years, is a mining entrepreneur, whereas Gyang Pam, Davou Pam and Musa Nura are his laborers.
36. Ezra Pam, interviewed 12/31/08.
37. A gangue is a mound created from heaps of earth removed from mined sites.
38. Interviews with Abubakar Yau, Ezra Pam, Gyang Pam, Musa Nura, Davou Pam, Davou Pam, Chuwang Dung, Idowu Fedelis, Emmanuel Boyi, Salisu Adamu, Ibrahim Gwom and Terry Ephraim.
39. Chung Dung, interviewed 01/10/09.
40. Ephraim Terry, interviewed 01/12/09.
41. Interview with Salisu Adamu, at Bisichi, 01/12/09.
42. S. U. Fwatshak, "Garnet mining and its impact on Jiblik communities from the 1980s to the 1990s" in *History Research and Methodology in Africa,* ed. Mike O. Odey, John G. Nengel and Okpeh Okpeh (Makurdi, Nigeria: Aboki Publishers, 2007), 271.
43. Ibid., 272.
44. Ibid.
45. My failed attempts to meet with Husseini were made when carrying out field work on garnet mining in March 1996 at A. M. Kalambu office in Dilimi street, Jos. This was during the heyday of garnet mining. I did not, however, record the specific dates of my visit to the company office.
46. This involved an arrangement between the landowners and the miners to the effect either that the miners paid a fixed rate of N50 each per day or agreed to share the proceeds of whatever quantity of Garnet they got from mining the land. Under the former arrangement, payment was up front and the landowner was responsible for reclaiming his land. Under the latter, the payment was made at the end of the day or at such a time any/all miner(s) affected by the agreement hit luck and sold his/her/their find. If at the end of the day the miner(s) found nothing, the landowner got no pay but the miner(s) was/were responsible for reclaiming the land. In both arrangements each landowner tried to closely supervise the operations of the people mining his land. There were also miners who were neither company workers nor people who had made any arrangements with landowners. They just mined any piece of land as if the land was free for all. This category of miners who were large in number continued to be a source of conflict between the landowners and the miners on one hand and between themselves (illegal miners) on the other. But the Companies did not seem to care about how Garnet was obtained, as they did about to whom it was sold.
47. Emmanuel J. Dasat, "The Need for a Sustainable Environment," in *Land Desecration and Environmental Degradation on Nigeria's Jos Plateau,* ed. Bagudu, 21; See also M. J. Alexander, "Soil Characteristics and the Factors Influencing their Development the Mine Spoil of the Jos Plateau," Jos Plateau Environmental Resources Development Programme, University Linkage Durham (UK)-Jos (Nigeria) Interim Report No. 11, August 1986, 126.
48. D. D. Dabi and J. L. Nyagba, "An Alternative Approach to the Reclamation of Mine Lands on the Jos Plateau, Nigeria," *Journal of Environmental Sciences* 3, no. 2: 245.
49. M. J. Alexander, "Soil Characteristics and the Factors Influencing their Development," 126.
50. Ibid.
51. Ibid.
52. John Dung Gwom, "The Impact of Tin Mining on Economic Activities in Plateau State," in *Land Desecration and Environmental Degradation on Nigeria's Jos Plateau,* ed. Bagudu, 42, 50–52.

53. Mangvwat, "Tin Mining and Environmental Degradation," 61–62.
54. Ibid., 63.
55. Amalgamated Tin Mines of Nigeria Limited, "The Plateau Minesfield: What you should know about Nigerian Tin Production" (n.d.), 6.
56. Much of the Mines Reclaimed Areas (MRAs) are planted with eucalyptus and are indicated on the mining sheets at Consolidated Tin Mines Headquarters, Bukuru.
57. Emmanuel J. Dasat, "The need for a Sustainable Environment," 24–25.
58. M. J. Alexander, "Soil Characteristics and the Factors Influencing their Development," 144.
59. S. U. Fwatshak, "Garnet Mining," 282.
60. *Minerals Act,* 1946, Part IV: Provisions Regarding Water.
61. Gonyok, "A History of Labour," 286–381.
62. Ibid., 385–394; Freund, *Capital and Labor,* 164.
63. Sam G. Colom, interviewed January 9, 2009. Consolidated has files on compensation going back to the 1960s. For example, in the 1970s, one Peter Shak was paid 62 pounds compensation for his land measuring 1.5 acres taken for mining.
64. Abubakar Ya'u, interviewed 12/31/08.
65. ATMN: "The Plateau Minesfield" (n.d.), 6.
66. Davou, "Tin Mining and the Ecology of Plateau State," 35; Mangvwat, "Tin Mining and Environmental Degradation," 67; Gwom, "The Impact of Tin Mining on Economic Activities in Plateau State," 53; Dasat, "The Need for a Sustainable Environment," 23.
67. Nankin Bagudu "Welcome Remarks," in *Land Desecration and Environmental Degradation on Nigeria's Jos Plateau,* ed. Bagudu, 10–12; Sam G. Colom, interviewed January 9, 2009.
68. Fwatshak, "Garnet Mining," 288.

3 Biological-Based Warfare in Central Nigeria

An Aspect of Pre-Colonial Military Science and Technology

Olayemi Akinwumi and Mailafiya Aruwa Filaba

INTRODUCTION

Three major factors informed the science and technology of traditional biological agents. The first one was the environmental wealth in terms of the availability of natural plants and big game animals that were difficult to hunt. The second factor was the ecological challenges related to disease. The third and perhaps the most dominant factor was violent conflicts in the form of external aggression including wars, raids and rivalries in the communities over access to land and wealth and disputes about traditional stools, wives or slaves. Field research included interviewees from the Gbagyi (Gwari) and Bassa in central Nigeria and Borgu and Bussa in Dahomey and in Nigeria. Interviews revealed that one of the secrets behind the survival of the Songhai, Jihadists, Ebira, Nupe and Yoruba sporadic raids was their possession of the knowledge of the science and technology of biological agents, the iron industry and hilltop refuge centers. The groups embarked on the secret production of armaments until they were forcibly stopped and seized by colonial conquest. They attributed mass production to the incessant attacks from warring groups and imperialists in their search for slaves for the Trans-Sahara and Trans-Atlantic Slave Trade and for slave-based agriculture and plantations to feed the Royal Niger Company (RNC). They still used the armaments for hunting big game and fishing.

Production of biological agents was an offshoot of medicine. Intensive study of the traditional medical industries and practitioners reveal that the practice was boosted in central Nigeria by the environmental availability of plants of all sorts, a few of which are deadly poisons harnessed for biological weapons and for the treatment of diseases. Whereas almost all the communities in central Nigeria have professionals in ethno-botany, treatment of diseases and the production of poisons, the Borgu, Bussa, Bassa and Gbagyi rated as the best and neighboring groups consulted them. It is expected of every adult among the Gbagyi to have some knowledge of herbs, which is the selection and use of plants (grass, shrubs, barks, seeds, flowers and saps), sand, dropping of animals, stones, bones, some animals, birds and assorted creatures for medicine. Thus, diseases enabled the people to develop ethno-botany. The people knew the common plants for the most common diseases and were quite efficacious, and which plants were additives, sedatives, stimulants, hypnotics, anti-pains, poisons and so on. There are known herbs and shrubs for particular diseases, as they have proven to be effective over millenniums. However, there was a particular class of armament production and poisoning professionals just as there was a class of warriors.[1]

PREDILECTION FOR WAR IN THE PRE-COLONIAL DAYS

The science and technology of biological weapons was precipitated by the incessant violent conflicts, slave raids, wars and dry-season hunting.[2] Wars were caused by many factors: massive migration and the struggle for land and the scramble for natural resources between the autochthonous communities and immigrants; the search for slaves for the slave trade; outbursts of sociopolitical struggle for headship or imperial domination; and quarrels over land, women and slaves. The 'lucrative' nature of the slave trade encouraged some syndicates and communities to embark on raids of weaker neighbors. The slave trade received impetus in Africa when the Roman Empire's demand increased for African slaves to build roads and estates in Rome as well as for domestic and farming labor. Arabs then started to trade for slaves from Northern and Eastern parts of Africa in the ninth century AD[3] and in central Nigeria dating from the fourteenth century. Slavery became the major item of intercontinental trade in the fifteenth century when Europeans initiated the Trans Atlantic Slave Trade.[4] Demand for slaves informed the barbarous internecine wars. Slave raiding peaked in the seventeenth, eighteenth and nineteenth centuries, and what is now central Nigeria suffered slave raids organized by the rulers of Nupe, Ebira, Igala and the Emirs and their syndicate adventurers like Nagwamatse and the Magajin Keffi Yamusa (whose raids earned him the title of "the master pillager"). They acquired European firearms with which to embark on wars to get captives for the trade. Of course, the emergence and survival of the Sokoto Caliphate was based on slavery and the slave trade, and at the eve of conquest it had about 2,000,000 slaves,[5] most of whom were from central Nigeria.

Most of the slaves particularly in Sokoto, Kano and Zaria, were from Nasarawa Province. Also, Ilorin and other Nupe areas like Bida, Lapai, Agae, Bajibo and Jeba had a large concentration of slaves, most of whom were from Gbagyi and Bassa and worked plantation (*tungazi*) agriculture that grew cotton, groundnuts and various other items to supply the Royal Niger Company.[6] Magajin Keffi Yamusa 'the master pillager' killed Capt.

Moloney to continue with his cherished headhunting in the neighborhood with organized violence, banditry and kidnapping. Violent enslavement was a continuum from large-scale political action and resulted in the continuous enslavement or re-enslavement of people.[7]

European travelers noted that the "raids on neighboring pagans were considered (by the Muslims) a pleasant and profitable way of spending the dry season," and "the methods adopted for this end (replenishment of their dissipated economy) were the wholesale plunder of the populace. Slaves were the general currency, and under the extensive system of slavery there was an absence of social cohesion, the mass of the subject pagan peoples being reduced to a state of complete economic and moral paralysis."[8] Oldfield similarly reported that the jihadist "travelled very quickly, took the unsuspected inhabitants by surprise and caught hundreds of prisoners as well as cattle and horses and disposed of them to the Arabs and slavers on the banks of the Niger, and eventually reached the seaside."[9] With the support of Arabs, the Fulani staged a successful coup in Nupeland in 1845. They also destroyed their rival brothers, and were known to behave in a cruel manner and were hated by neighboring groups. They raided and sold captives into slavery, destroyed settlements and took properties as booty. Thus, others suspected the Fulani to be on a bad course and as the "most barbarous race."[10] The Nassarawa people had difficulty in conquering the Gbagyi and Bassa and their neighbors because of the biological weapons in their possession. This difficulty can be inferred from the request the ruler of Nassarawa made to the RNC, as he pleaded for rifles and told MacDonald that he and his people were "not tillers of the soil; they did not dig the ground neither did they parter wory [bribe or pay a settlement] or palm the oil like the merchants . . . no; they were fighting people; they fought the pagans and made slaves of them."[11] Sultan Bello similarly wrote a threat letter to ruler of Opanda in 1824, which stressed that non-Muslim rulers must surrender their territories to Muslim rulers and traders or face enslavement and destruction.[12] Indeed, the triangular slave trade and the establishment of the Sokoto Caliphate intensified the introduction of the slave mode of production in Hausa cities and among the Nupe and Ebira capitals. The urbanization of Hausa cities was partly characterized by a dialectical relationship whereby the suburbs and non-Hausa groups south of Hausaland were raided. Some extant traditions in Hausa cities corroborate this view, and claim that most of the slave estates (Rinji, Rumada, Gandu, Murgu) in Zaria in the nineteenth century were full of non-Muslim Gwari. The rulers of Sokoto asked the Emirs to send to them slaves and booty in the 1850s, as the economy of the Emirates was dependent on slavery.[13]

On occasion, Muslims enslaved their kinsmen. Ahmad Baba observed in the seventeenth century that the rulers of Hausa cities were Muslims who sometimes fought wars and took as many Muslim prisoners as they could and sold them out, just as they bought slaves whose backgrounds were non-Muslim.[14] In the Sokoto Caliphate in the nineteenth century, Emirates raided Emirates for slaves, and the Caliph imposed taxes in slaves on the Emirates. Zaria in particular raided the Emirates down to the Niger-Benue confluence. In the last three decades of the nineteenth century, Zaria raided the suburbs of Keffi, Abuja and Nasarawa. Also, Keffi and Abuja fought each other and as well fell out with the Emir of Jema'an Darroro over boundary issues.[15] The incessant raids of the region by Zaria made Magajin Keffi to stop Keffi from being turbaned from Zaria. The demographic consequences of the raids can be inferred from this remark:

> About the beginning of the 19th century the Fulani invasion commenced in earnest. . . . It was not until about the end of the 18th century that the more Northern District of Nassarawa Province began to suffer from slave-raiders and that parties of Fulani and Habe commenced to settle in them. The first ten years of the 19th century saw certain Fulani and Habe family established in fortified strongholds from whence they raided the surrounding country, and sent northwards long caravans of slaves. They penetrated southwards and even crossed the Benue and raided in the Idah Kingdom. The end of the 19th century saw a complete ruins and depopulation of the Province. . . .
>
> In Nassarawa country, a once fertile and populous province, one can only see the remains and ruins of large and totally deserted towns, bearing witness to the desolation rough by hundred years of internecine strife and slave raiding by the Fulani.
>
> The remaining inhabitants of much [*sic*] towns fled to the hills in all directions. Those who approached the Eastern and North-Eastern mountains of the Province, until they learned how to defend themselves, were further raided by the Headhunting tribes who inhabit these hilly localities. . . . Such was the state of the Province when the arrival of Sir Frederick Lugard put a stop to the slave-raiding and evolved law and order out of chaos and ruin.[16]

That was why "slavery figured prominently in the (colonial) conquest itself."[17]

The dialectical effect of the slave raids was that the Gbagyi lost their population, whereas those of the Hausa, Kanuri, Nupe, Ebira and the coastal states' population grew. With the establishment of Keffi and Nasarawa towns, the Gbagyi confederation of Kurudu, Karu and Kurape kingdoms shrank by losing those territories to the Zaria Emirate, although the capitals and settlements of about twenty kilometers radius were not captured and incorporated into the Emirate.[18]

Similarly, wars in Borgu area and beyond are reported in by both primary and secondary sources, some of which were surmised by Olayemi Akinwumi.[19] Tradition indicates that the ruling class, the *Wasangari*, was a warrior class that invaded the country on horseback with military weapons. This explains why the princes formed the backbone of the entire military contingent in the country. Qualification to the highest position in any of the states in Borgu depended much on the military prowess of every prince. Warfare in Borgu was therefore the responsibility of the ruling class. The rulers provided the leadership and the weapons to the other classes or groups who formed the bulk of the archers and infantry. This actually formed the strength of the Borgu army. In the 1836 Ilorin–Borgu war, tradition indicates that the Borgu provided about 400 horses, as it did with other wars.[20] Warfare was a serious business in Borgu. To an average Borgu, a military defeat meant death. A Borgawa never allowed himself to be enslaved and therefore did anything to win, even if the war was prolonged, or the country was under siege. The Borgawa soldiers, therefore, relied on their weapons and biological agents for the greatly desired victory.[21]

Some of the weapons at the disposal of the army included bows and arrows. In fact, they were the primary and most useful weapons in the country. The infantry soldiers who formed the bulk of the army were all armed. Each archer normally carried up to forty or fifty arrows kept in a leather or bamboo quiver, hung on the individual's shoulder. Usually, these arrows were not exhausted before the cavalry took over, and when it did, it was always decisive. Another weapon available to Borgawa soldiers was the spear (*sari; mashi; masai*), used by the cavalry. It was used to engage the enemies at a greater distance. A typical *sari* was made of an iron-head fixed to a longer shaft of about five feet. Usually the *sari* were produced in large quantities in time of peace and safely kept in the armory, and in the case of Busa, at the king's palace. Among the Gbagyi they were stored in the town sanctuary (*apolo/adawyiyagbe*) or with the commander of the army called *Bezye yaako/Barde,* the custodian of the armory. Other weapons of lesser significance were swords (*flenida; takobi; obe-yako*) and hand axes (*gban; gatari; gasasawa/aze*). These weapons were utilized in close hand-to-hand fighting. In the case of the sword, it was probably introduced to Borguland from Hausaland. Weapons used by other communities in central Nigeria included wooden spears, stone-pebble bullets thrown with slings (*gbataaknu*), sticks; knives; iron traps; wooden and rope traps; horns; bones and den-gun.[22] All these weapons were used for the defense of the country, and they all proved effective. For example, communities survived the invasion of Songhai in the sixteenth century when it was the then most powerful military state in Sub-Saharan Africa under Sonni Ali and later Askia Mohammed and the nineteenth-century's series of wars and slave raids because of the effective use of local weapons.[23]

Dahomey under King Tegbesu became a powerful military state that relied mostly on its cavalry, muskets, discipline and use of famous women soldiers in the Amazon. Mockley-Ferryman travelled up the Niger in 1892 and noted, "their (*Borgawa*) poisoned arrows enabled them to hold their own with the forces of Dahomey notwithstanding the latter's muskets." Similarly, the Habe rulers' menace was curtailed by Borgu. The various Habe rulers were in the process of establishing strong states, and Borgu being very close was a targeted-satellite state always attacked by Bussa and Kebbi, but Borgu was always victorious. One of the biggest attacks came when Sheikh Usman dan Fodio instructed all of his faithful followers to launch attacks on their neighbors. This led to the conquest of Nupeland and Ilorin Malam Dendo and Alimi. They then attacked Borgu determined to bring down its 'pagan' administration. Borgu, then under the leadership of Nikki, succeeded in repulsing the attacks and remained independent of the jihadists.[24]

HARNESSING VENOM AND POISONS FOR BIOLOGICAL WARFARE

This section will discuss the professionals who made biological agents and the various sources of the poisons. Communities that engaged in biological agents had trained professionals. In Borgu, there were different classes or groups of people involved. In Bussa, the *Larus* were responsible for the weapons and the biological agents. Among the Bokos of Nigeria and Benin Republic, especially at Pusisa, the hunters were responsible. Among the Batonu of Nigeria and Benin Republic, a class of medicine men was responsible. Among the Bassa and Gbagyi, it was the *azakwoyi* (elders), *abeyi* (priests) and *Bezye* (army commanders), *pya-dalpyagbe* (blacksmith) and *baushe* (chief-hunter). However, not all the *Larus,* hunters and medicine men there knew the secrets of biological agents. In Bussa, the *Laru* of Shagamu were responsible for this, whereas in Kaiama, the Ekana people were responsible. Yina and Bouanri in Benin Republic along with Kenu, Gure and Moshe in Nigeria were the most popular areas in Batonu where poisoning was practiced. Across Borgu people visited these areas to have their weapons poisoned. Thus, the few with the knowledge of biological agents were known. Borgawa warriors were aware that preparation for war was not complete until the biological agents were duly applied to the weapons.[25]

Preparation of biological agents differed from one community to another. Among the Batonu, it was more ritualistic. The people involved commenced fasting for three days before the preparation started and abstained from sex for twenty-one days. During the preparation, their only dress was *san-roko* skin tied around their waists. This indulgence was believed to make the agents effective. Among the Gbagyi, the elders (*azakwoi*) and *pyagbe* (blacksmith) withdrew into the forest or top hill for the craft.[26]

The class that produced the armaments mystified their positions. Everything is mixed up with myths. Of course, some of these professionals were priests or soothsayers. Some of them claimed to be magicians or wizards. They claimed sorcery, mysticism and magic and ordered bees to sting only the enemies even in a mixed crowd. They could send snakes to identify their enemies and bite them. They caused their enemies to die, only to see them again as a slave in the neighboring village. This is called *nuwna-lo* among the Gbagyi, and as *dabon-kwotto* in Hausa.[27]

There were many sources of biological agents. There were poisons derived from reptiles, animals, chemicals, plants and animals. The most common of all was the agent obtained from a snake's saliva, teeth and head. Snakes that provided venom include vipers, cobras and adders. Venom was also obtained from toads' sweat. Any toad bitten by a snake acquired snake venom. When bitten by a snake or shot with a poisoned arrow and the victim survives, the sickness recurs every year at the exact period the incident happened. Laboratory tests and biochemical analysis of the poisons, particularly the venom derived from snakes, reveal that the venoms are organic substances that are a biodegradable poison that cannot be digested. Whenever the venom gets into a foreign system (someone's body), it gets amalgamated to the blood cell and engulfs it. This covers the blood cells in ferrocene that carries oxygen and prevents it from getting oxygen. It then exchanges the oxygen the blood should get with an organic substance. This causes a rupture of the cell membrane, particularly the white blood cell. The venom continues to travel along the circulatory system fighting the white blood cells, and the immune system is weakened. This makes other body organs surrender to the venom. On reaching the nervous system, the venom causes more damage that leads to death.[28]

Venom was collected from bees (*osu*) and wasps (*muiznu*), which caused pain and swelling. Dogs suffering from rabies were killed, and their heads dried and pounded. The powder was applied on prisoners of war to cause madness, and on fighting instruments so that the victims later became mad.[29] Poison was also derived from the heart of porcupines. This creature consumes poisonous insects and plants and the poison accumulates in its liver and heart, which when eaten, can cause death. Similarly, the mucus and horns of gazelle and buffalo contain poison, which prevents wounds from healing.[30]

Thus, when shot by poisoned arrows, or struck with a spear or bruised with a poisoned object, one experienced an instant shock as the motor sensory nerves sent a message to the central nervous system. The victim experienced excessive sedation (sleepiness and weakness). The victim could not walk up to ten meters before falling down when shot with poisoned arrows or spears. Resistance to the venom depended on the individual's immune systems.[31] Poisons used in treating war weapons, usually arrows and spears, could be active or effective for fifty years.[32]

Some of the plants with high toxin concentrations include *strophantus hispidus, erithrophlem guineense* and *euphobia hirta.* These plants were known to contain poisons that could either paralyze or lead to instant death. *Strophantus hipidus,* called *dea,* was locally grown in the country. The plant is especially known to exist in rocky places. Its seeds were needed for poisons. For example, it contains glycoside mixture, K-strophanthin, alkaloid trigonelline, fatty oil, resin and enzyme, which when used as poison could lead to paralysis of the heart and then death within thirty minutes.[33] *Obwa* (soya bean like plant but with bigger beans) seeds were ground and boiled till the water thickens.[34] Cactus plants contained poisonous milk, which caused instant rashes and burns of skin. *Obwaa*, a tobacco-like plant, leafless evergreen milky plants, contained poison; hyacinth and flower and sugar-like grass and roots caused heart failure or asthma. Poison was also derived from mushrooms and shrubs/herbs. Their contents are fungi-like, which cannot be destroyed by white cells.[35]

Other poisons include itchy substances. The most efficacious one is the *karara/bugyeesnu*—brownish dust/hair of a creeping plant mostly in rivers and forests. The dust contains histamine, which when contacted stimulates the skin causing sever itching. White blood cells attempt to fight the histamine, but as the white blood cells cannot get to the surface of the skin to fight it, the skin would form nodules, rashes or pimples.[36] *Zakami* (a garden-egg-like flower) seeds contained high amphetamines. It could be soaked and the water was added to drinks to cause overjoy and the enemy would say 'all the secrets' when he or she drank it. The powder when taken in excess of one milligram can cause hysteria, trance and unconsciousness.[37] Ground pepper was used to torture war captives or for poisoning animals. Ground pepper was poured on dusty roads for cows to inhale and develop incurable pneumonia and catarrh potent enough to cause their death.[38] *Kwaya*, a brownish chemical stone found in rivers, contains metallic iron that is not degradable and when consumed can cause death. *Gizaka/Nama-nama* is a deadly hairy worm that causes instant big rashes and burns of the skin.[39] Some scorpion (*napisis*) stings were used to poison foods and drinks of enemies so that they died of ulcers and blisters.[40] Although one or two of these were combined for hunting big game, all of them or more deadly ones were processed together for war.[41]

Chameleon's dung when poured on weapons was believed to break the victims' ability to be bulletproof and iron proof. Gun bullets, arrows, spears, knives, sticks, whips, pebbles and other things to be used on enemies were poisoned.[42] Indeed, warfare was more of a business in which all strategies available to the combating parties were utilized to gain victory. In this case, local lethal weapons were produced or imported, and defensive mechanisms, ditches and fences were built and biological techniques introduced to defend their sovereignty. Through the innovation of reliable military technology and the use of biological agents, Borgu and Gbagyi states were able to maintain their pre-colonial territorial integrity in spite of the fact that they were surrounded by powerful imperialistic states like Songhai, Oyo, Nupe and Zaria and later the Sokoto Caliphate. Almost all communities in Africa used biological weapons against their opponents. The biological agents in this sense meant arrow poisoning, which was the oldest method in the continent. Also, there was poisoning of enemies' horses and water supply. When this was done, the opposing camp not only suffered military defeat, but also became incorporated as satellite states into the victorious country.[43]

Assorted ingredients were dried, crushed and later mixed with other ingredients. This concoction was smeared on arrows and spears and then left to dry. The object immediately became dangerous and could lead to instant death if it wounded someone, whereas toxins from animals and plants were mixed up together and smeared on weapons. This was often done by a special class of people and not ordinary members of the society.[44] In the case of Borgu, the preparation of poisons varied from state to state. The most popular method was to combine together both the toxins from plants and animals. The act of poisoning was known as *sewe/omwui-du,* which could include one ingredient or many depending on how efficacious they wanted it and for what purpose.[45]

To prepare the poisons, the desired ingredients were first collated. Some of them were boiled or first soaked in cold water before boiling them in a big pot. The heating was done by warming it with hot charcoals and not firewood and flame. Some of them were crushed or ground and their sap or powder were mixed with other concoctions before the arrows were dipped in, or smeared with, the poison. Sacrifices were made and incantations were chanted to appease the spirits.[46] the boiling process, the mixture was filtered and as many as one thousand arrows could be smeared with the poison.[47]

Borgu biological agents were not used for destruction only, but they were also used to hypnotize their enemies and to make the warriors bold and to act as anti-poison (antidote) for poisoned weapons of the enemies. To hypnotize the enemy, the Batonu of Borgu used a powdery substance that was spread at the war front. If inhaled by the enemies, it had the effect of taking the enemies into a stupor, becoming drowsy or sleepy. When this was done the Borguwa would easily capture and enslave the victims.[48]

To have the knowledge of the agents for biological warfare was one aspect; there was also the aspect of knowing the antidotes. Among the Batonu of Borgu, antidote is called *bwani,* and among Gbagyi as *omwui-shigbe.* The professionals also produced the antidotes. The antidote for poisoned arrows were concoctions of *ash-water* (alkali—sodium hydroxide [NaOH]), natron, bitter lemon, sour ingredients (acids) like *ama/yakuwa/zobo* and so on. Some of the antidotes such as anti-snake (*wa-sa*) were sap/backs of *jiyi/dinya* (vitex-doniana), *gbakopiyi/gwandan-daji,* shea-butter tree (butyrospermum paradoxum), *myimyi/kokiya* (strychnos spinesa), *jemui/lemun-tsami* (carissa edulis), red oil, soap, *oznu*—palm trees (elaeis guineensis) and so on. Antidotes were applied orally on the victim to neutralize the effect of the poison in his body. The colonial administrators noted the efficacy of the antidotes, as the Kaiama people treated Lord Lugard with such an antidote when he was attacked with poisoned weapons on his way to Nikki to sign a treaty with the king. Eventually when he became the Governor-General of Nigeria in 1900, he compensated Kaiama by making it an Emirate and at the same time the capital of the newly created Borgu Province.[49]

It will be important at this juncture to identify some of the major weaknesses of the production of biological agents as well as submit suggestions on how to better production. Production of biological agents was done by the same class of people responsible for general herbal medicines. The priests/priestesses kept their knowledge secret. They even covered the herbs with a canopy of ropes, leaves or cotton so that the client might not know the content. Professionals did so to keep a monopoly. Secrecy included lack of training people outside the family so that the knowledge of medicine remained only in the family. Many people and families passed away with their knowledge, which became extinct as the practitioners were illiterates who did not write down their arts and knowledge for posterity. Secrecy was characterized by jealousy and lack of joint efforts and conferences through which they could exchange ideas. Very few herbalists knew or informed clients on the side effects of the herb antidotes.[50] It was also discovered that there was inadequate regulation of created biological agents, a lack of scientific studies on safety and effectiveness and possible harmful interactions with other medicines.[51] There was an issue of mystifying everything by the priests and priestesses, which most of the time eluded a proper understanding of the causes of afflictions. There was also the danger of contamination and dirt. Although they had antidotes, it was not certain that they completely neutralized the venom. Priests rarely tended botanical gardens but lived at the mercy of the natural environment.[52]

The technology and science of the production of biological agents in pre-colonial Africa was encouraged by the incessant internecine wars. The scientific knowledge involved ethno-botany, blacksmithing and biochemistry by using ingredients abundant in the region. The frequency of internecine wars indeed informed the mass production of arrows and spears treated with poisons to render them deadly, which were used to safeguard individuals, communities and the territorial integrity of kingdoms. The Gbagyi and Borgu were consulted owing to their knowledge of making poisons. As evidence of this aptitude, these small groups successfully safeguarded and maintained their independence throughout the pre-colonial period. Also, traditional biological agents included antidotes for the destructive weapons. This also involved research into the use of the right plants to counter the destructive effects of the biological agents in the body chemistry. Africa is encapsulated in violent conflicts that lead to sporadic outbursts and the antagonistic groups still apply traditional biological agents on each other; hence, there is the need to study them and their antidotes in depth for mass production. Modern pharmaceutical companies and institutions should contact herbalists for any input and vice versa. The very ingredients for the biological agents should undergo thorough laboratory analysis. Pharmacists should organize conferences for herbalists. Laboratories' input is a desideratum to know about the safety and effectiveness of herbal remedies.

"Survey Report on Biological Weapons amongst the Borgu, Bussa and Gbagyi People in July 1992 and Dec. 1993." The resource persons were Ba-Karabonde Sabi in New Busa on July 6, 1992 *Arewa House World Conference on*

Slavery and society in History; Hassan Gbesitaku in Kaima, on December 5, 1993; Garuba Woru, New Bussa on December 12, 1993; Mallam Sabi in Ilorin on December 12, 1993; Maurat Kissira on July 16, 1992; Samila Yaru on December 2, 1993. Other resource persons contacted by M. A. Filaba included Mr. Danjuma Shagye, a Chemist at NSUK Keffi in April 2008 and February 2009; and Rt. Sgt. Dauda Kaka Kure at Nasarawa in April 2008 and February 2009. We acknowledge the Faculty of Physical Sciences NSUK for making available to us the laboratories and the staff for experiments.

NOTES

1. M. A. Filaba, "Gbagyi Traditional Disease Diagnosis and Treatment: Problems and Prospects," in O. Akinwumi, O. Okpeh, C. Ogbogbo and O. Adoyi, *African Indigenous Science and Knowledge System* (Nigeria, Abuja: Roots Books and Journals Ltd., 2007), 378–397.
2. M. A. Filaba, "Slavery and Slave Trade: Ageless and Nebulous Syndrome of Servitude and Commerce in Central Nigeria," in *The Global South 2.2 (August 2008): Hegemony and Subalternity: Africa in a Global Age. The Global South,* ed. Ashok K. Mohapatra, Alfred J. López and Adetayo Alabi, at http://inscribe.iupress.org/loi/gso. Accessed October 2008.
3. Donald R. Wright, "Slavery in Africa," in *Microsoft ® Encarta ® Encyclopedia 2002.* © 1993–2002 Microsoft Corp.
4. See papers in A. Mahadi, ed., *Arewa House World Conference on Slavery and society in History.* Arewa House Kaduna, March 26–30, 1990. The following papers elaborately discussed the issues: A. Mahadi, "Perpetration of the Concept and Institution of Slavery and other forms of Violence in the name of Islam: The Case of Bilad As Sudan to the end of the 19th c"; E. Oyedele, "Slavery and Urbanization in Central Sudan up to 1900 A.D."; M. Y. Mangvat, "Slaves and Slavery on the Jos Plateau in the 11th century: A Review of Evidence"; E. Naley, "The Stealing of African to the United States and the perpetuation of European Cultural Hegemony"; C. N. Ubah, "Slave Trade in Africa on the Historical Screen"; C. N. Ubah, "Slavery and the Emancipation in the Nigerian Emirates"; J. E. Philips, "Slavery and Ribats in the Sokoto Caliphate especially in Comparison with Slavery in the United States"; B. Achi, "Warfare and Slavery in Zaria Emirate in the 19th c"; I. M. Jumare, "Slave Agricultural Estates in the Sokoto Caliphate"; and R. A. Goodridge, "Muslims and Non-Muslims: The Issue of 'Slavery' in the Establishment of British Rule in Northern Cameroons to 1927."
5. Paul E. Lovejoy and Jan S. Hogendorn, *Slow Death for Slavery: The Course of Abolition in Northern Nigeria, 1897–1936* (Cambridge: Cambridge University Press, 1997), 32.
6. Ibid., 34–35, 49–52; S. Mohammed, "Gbagyi and their South West Neighbors to 1898," in *Studies in the History of Central Nigeria Area: Volume 1,* ed. Idrees and Ochefu (Lagos: CSS Ltd, 2002), 513–532.
7. Ibid., Chapters 1 and 2; J. R. Willis, *Gazetteers of Northern Provinces of Nigeria, The Central Kingdoms, vol. 3* (London: Frank Cass, 1920); "Notes on Nassarawa Province," 3, 19.
8. C. K. Meek, *The Northern Tribes of Nigeria: An Ethnographical Account of the Northern Provinces & Report of 1921 Decennial Census,* vol. 1 (New York: Routledge, 1971), 257–258, 287–293. Also see Philip D. Curtin, "The

Atlantic Slave Trade 1600–1800," in *History of West Africa Vol. 1,* ed. A. J. Ajayi and M. Crowder (New York: Columbia University Press, 1974), 240–242.
9. MacGregor Laird and R. A. K. Oldfield, *Narrative of an Expedition into the Interior of Africa by the River Niger in 1832, 1833 & 1834, Vol. 1* (London: Frank Cass, 1967), 90.
10. W. B. Baikie, *Narrative of an Exploration Voyage up the River Kwora and Binue in 1854* (London: Frank Cass & Co Ltd., 1966), 270.
11. C. M. MacDonald, "Exploration of the Benue," in *Journal of the Historical Society of Nigeria* 1, no. 3 (1958): 512. J. A. Ohiare, "The Kingdom of Igu and Opanda, c. 1700–1939: A Study in Intergroup Relations" (PhD thesis, A.AB.U. Zaria, 1987), 167.
12. R. Lander, *Records of Capt. Clapperton's Last Expedition to Africa,* 2 vol. (London: Frank Cass, 1830; repr. 1967), 47–55.
13. E. Oyedele, "Slavery and Urbanization in Central Sudan up to 1900 A.D.," in *Arewa House World Conference on Slavery and Society in History,* ed. A. Mahadi (Kaduna, Nigeria: Arewa House, 1990), 8–9; J. Ohiare, "The Kingdom of Igu and Opanda, c. 1700–1939: A Study in Intergroup Relations" (PhD thesis, A.AB.U. Zaria, 1988), 400–450.
14. Ahmad Baba "Slavery and Islam in the 19th century Central Sudan," in *Nigerian Perspectives: An Historical Anthology,* ed. T. Hodgkin (London: Oxford University Press, 1975), 154–155.
15. T. Sciortinor, "Notes on Nasarawa Province," in *Gazetteers of Northern Provinces of Nigeria,* ed. J. Willis, Kirk-Green (London: Frank Cass & Co., 1920), 9, 10, 15; M. G. Smith, *Government in Zazzau,* 2nd ed. (London: Frank Cass & Co., 1978), 3, 8; O. Temple, ed., *Notes on the Tribes, Provinces, Emirates of Northern Nigeria* (London: Frank Cass, 1965), 152; M. A. Filaba, *Kurape, Karu. Kurudu and Rubochi Kingdoms: Gbagyi Confederation in Central Nigeria to 1960* (Ethiopia: Jimma University Press, 2007), 161–162.
16. Sciortino, "Notes on Nasarawa Province," 15.
17. Lovejoy and Hogendorn, *Slow Death for Slavery,* 1.
18. Filaba, *Kurape, Karu. Kurudu and Rubochi Kingdoms,* 158–204.
19. Olayemi Akinwumi, "Biological-Based Warfare in the Pre-colonial Borgu Society of Nigeria and Republic of Benin," *Transafrican Journal of History* 24 (1995): 123–130.
20. Ibid.
21. Ibid.
22. Ibid.; Filaba, "Slavery and Slave Trade," 158–203.
23. Ibid.
24. J. F. Ade Ajayi and Robert Smith, *Yoruba Warfare in the 19th Century* (Cambridge: Cambridge University Press, 1971); O. D. Akinwumi, "The Oyo Borgu Military Alliance of 1935: A Case-study in the Pre-colonial Military History," *Transafrican Journal of History* 21 (1992): 159–170; J. C. Anene, "The Eclipse of the Borgawa," *Journal of the Historical Society of Nigeria* 3, no. 2 (1965): 211–218; J. C. Anene, *The International Boundaries of Nigeria* (London: Longman, 1970); A. Bala, "Biologically-Based Warfare in Pre-colonial Nigeria," in *Science and Technology in African History with Case Studies from Nigeria, Sierra Leone, Zimbawe and Zambia,* ed. Gloria T. Emeagwali (New York: The Edwin Mollen Press, 1991); Michael Crowder, *Revolt in Bussa* (London, Faber and Faber, 1973); J. Echenberg, "Late Nineteenth Century Military Technology in upper Volter," *Journal of African History,* 12, no. 2 (1971): 241–254; M. B. Idrsi, "Political and Economic Relations in the Bariba State," (PhD thesis, Birmingham University, 1973);

R. Smith, "Yoruba Armament." *Journal of African History* 3, no. 1 (1967): 87–98; National Archives Kaduna (NAK/SNP 7/8 1858/1907); Survey in July 1992, Dec. 1993, and January 2009.

25. We are grateful to Mr. Danjuma Shagye, a chemist, Nasarawa State University, Keffi, for the biochemical analysis of the venom and its effects in foreign bodies.
26. Filaba, "Slavery and Slave Trade," 158–203.
27. Ibid.
28. Interview with Mr. Danjuma Shagye, a Chemist, Nasarawa State University, Keffi, for the biochemical analysis of the venom and its effects in foreign bodies, in April, 2008
29. Interviews with Mr. Danjuma Shagye and Rt. Sgt. Dauda Kaka in April 2008.
30. Ibid.
31. Ibid.
32. Ibid.
33. Ibid.
34. Ibid.
35. Ibid.
36. Ibid.
37. Ibid.
38. Ibid.
39. Ibid.
40. Ibid.
41. Ibid.
42. Interview with Aruwa Maishanu, a hunter, in April 2008.
43. Akinwumi, "Biological-Based Warfare in the Pre-colonial Borgu Society," 123–130; Filaba, "Slavery and Slave Trade," 158–203.
44. Ibid.
45. Akinwumi, "Biological-Based Warfare in the Pre-colonial Borgu Society," 123–130.
46. Ibid.
47. Ibid.
48. Ibid.
49. Ibid.
50. Filaba, "Gbagyi Traditional Disease Diagnosis and Treatment," 378–397.
51. *Microsoft Encarta Encyclopedia,* 2006.
52. Filaba, "Gbagyi Traditional Disease Diagnosis and Treatment," 378–397.

4 Religion, Science and Conflict

Problems for a Sustainable Environment for Human and Economic Development

T. T. Asojo and S. O. Ijaola

INTRODUCTION

> Has science made religion intellectually implausible? Does it rule out the existence of a personal God? Doesn't evolution, for instance, make the whole idea of divine providence incredible? And hasn't recent biology shown that life and mind are reducible to chemistry, thus rendering illusory the notions of soul and spirit? Need we any longer hold that the world is created by God? Or that we are really created by God? Or that we are really intended by something or someone to be here? Isn't it possible that all the complex pattering in nature is simply a product of blind chance? In an age of science can we honestly believe that there is nothing like direction or purpose to the universe? Moreover, isn't religion responsible for the ecological crisis. These questions make up the so called "problem" of science and religion. Today they may seem no closer to resolution than ever; yet they remain very much alive and continue to evoke an interesting range of responses.[1]

The aforementioned questions by J. F. Haught can be extended to issues of afterlife and the moral framework of society. We can also ask whether the idea of an afterlife is an illusion and do all religious dogmas suffer from illusions because man is reducible to breakable atoms? Does man have any need to obey a set of religious ethical rules because the idea of an afterlife remains illusory to modern material scientists? Are men not free to invent with their intellectual skills whatsoever they desire irrespective of its effect on their neighbors? Is all of life therefore about the survival of the fittest? However, *a fortiori,* there is a necessity to know if all the differences between science and religion are of any significant influence to human development in terms of developing a sustainable environment and economy in the developing world and in particular Africa where diverse religions have a strong influence on the people, informing both life and culture.

It is apparent that Africa today is still a developing continent, and this could be traced significantly to the effect of the conflict between religion and science owing to the imposition of European colonization and imported Western religion (Christianity) on her pre-colonial scientific worldview of the oneness of humanity and cosmos, which was unfortunately shattered in the minds Africans by secularism—a tool to separate religion from dictating people's life and most importantly to allow unhindered acceptability of value free modern science theories and technologies suffering from scrutinizing religious dogmas. Consequently, most Africans were left to struggle between their religion and culture and Western religions with their new ways of life and also between Africa's perspective of developing science and technology and the novel Western ones. Sadly, Africa's option was to remain a consumer of foreign technology by the system of colonialism and contemporary neo-colonialism. Hence, she has to work, but unfortunately to earn less, and her status as debtor in the international market is just a marginal adjustment, having to sell her raw materials cheap for expensive refined goods. At the core was the conflict between indigenous religion, an age-long heritage and the guiding principle for life, and modern science, which was stemming high when Western religion, colonialism and the slave trade held her bound.

The model of conflict between religion and science began to gain considerable attention in 1874 and 1896 through the writings of J. W. Draper and A. D. White.[2] Their emphasis created a model that lucidly identified the separation of two paramount spheres obviously inevitable to the survival of existential man:

> Religion is an aspect of the superstructure of a society, deriving ultimately from the degree of control and understanding of the material world. However, when man thinks in religious terms, he starts from the ideal rather than with the material world (which is beyond his comprehension). This creates a non-scientific and metaphysical way of viewing the world, which often conflicts with the scientific materialist outlook and with the development of society.[3]

Barbour observed that "the separation of the spheres of science and religion has been reinforced by the views of many scientists that science provides technical knowledge of a specialized kind, rather than a total philosophy of life,"[4] which religion tends to provide. This position is also held by Nobel laurel scientist Albert Einstein that "during the last century, and part of the one before, it was widely held that there was an irreconcilable conflict between knowledge (science) and belief (religion)."[5] In the same perspective, the nineteenth-century scholars J. W. Draper and A. D. White gave "a list of scientific explanations which overtly challenge religious sensibilities, and which ecclesiastics invariably protested at the presumption, and in which the scientists obviously would have the last laugh."[6] One such event was the hesitation of Churches to attach lightening rods.

> In 1745 the bell tower of St. Mark in Venice had once again been shattered in a storm. Within ten years, Benjamin Franklin had mastered the electrical nature of lightning. His conducting rod could have saved many a church from divine voice of rebuke, which thunder had often been supposed to be. But White reported that such meddling with providence, such presumption in controlling the artillery of heaven, was opposed so long by clerical authorities that the tower of St. Mark's was smitten again in 1761 and 1762. Not until 1766 was the conductor fixed—after which the monument was spared.[7]

The human and economic development of any nation is said to be proportional to its scientific horizon.

> A society develops economically as its members increase jointly their capacity for dealing with the environment. This capacity of dealing with the environment is dependent on the extent they understand the laws of nature (science), on the extent to which they put that understanding in to practice by developing tools (technology).[8]

It is therefore expected that conflict between science and religion, particularly among Africans where diverse religions flourish and dictate their philosophy of life,[9] would be a terrible setback to both human and economic development especially when science was to the advantage of the capitalist who championed the course of colonialism. Hence, this chapter aims at examining how the conflict between religion and science influenced the growth of human and economic development in Africa.

The following chapter examines the historicity of religion and science conflict, Africa in the conflict between religion and science, its effect on human and economic development and steps toward reconciling the conflict between religion and science. This study refers to science as natural science and analyzes science as a form of knowledge and technology with its affect on human thought and society. Religion is defined as a belief system that promotes theism with attributed moral values and eschatological inclinations. Christianity, Islam and African religions are considered in this study as the major relevant religions. The following sections focus on the conflict created in African societies between religion and science as Western religions imported the European-centered conflict between science and religion into Africa. Furthermore, Africa's colonial relationship with the West led to specific problems between science and the economy owing to the difference in African concepts about religion and science. The chapter finishes with a paradox between the desire of Africans to benefit from Western materialism but also the ethical problems they pose as well as the subaltern relationship between Africa and the West in terms of economic power and development.

HISTORICITY OF RELIGION AND SCIENCE CONFLICT

Humans are at the receiving end of religion and at the same time at the center of science; therefore, whatever the character of the relationship that ensued as a result of their interactions consequently cannot excuse human responsibility to sustainable environmental and economical development. Yet the experience of conflict between religion and science appear to be conspicuously historically unavoidable. The affects of this conflict on African communities led to far different results than the effects on their European counterparts. History provides the context in which to understand the conflict in Africa between religion, science and the environment.

Historically religion provided the vanguard of the body of knowledge that shaped the life of most societies before the rise of modern science in the fifteenth century. For instance "Hinduism is the philosophy, the religion, and culture of Indian people."[10] Every nation of the world seemed to embrace religion as the frame for their cultural development. It has been further observed that

> Anthropologists studying both tribal and world religions conclude that most religions include at least some explanations of natural phenomena. Sickness, death, meteorological phenomena such as rain or tornadoes, the existence and location of mountains and other land-forms, earthquakes, volcanoes, the passage of seasons and the positions of the sun, stars and planets frequently have religiously based explanations.[11]

This is not uncommon with African Religion as it also provides explanations about the natural world just like Christianity and Islam. The conflict between religion and science has been described as a large-scale war owing to "the famous Galileo incident, often portrayed as a contest between Catholic hierarchy (viewed) representing the forces of repression and tradition, the voice of the world, the dead hand of the past, and, the dulcet voice of reason and science."[12] It was also observed that the reigning Aristotelian concept introduced to astronomy gave rise to disputes between ecclesiastics and scientists:[13]

> While the Newtonian mechanistic world view tried to exclude all natural theology from science . . . and seek to protect theology from contamination by science. The metaphysical mixing of Science and religion, Boyle and Newton believed, corrupted true religion. . . . The gradual disengagement of Religion from explanations of natural world was well advanced by Darwin's day.[14]

The contemporary theory of evolution remained at logger heads with the theistic account of the creation story of Genesis Chapters 1 and 2. On the other hand the theistic argument that man and other creations were designed by God has been seen by modern evolutionists such as Richard Dawkins and Daniel Dennett as unacceptable. They affirmed that "human beings have not been designed, but are products of unguided blind process of natural selection operating on some such source of genetic variation and random genetic mutation."[15]

Another area of conflict between religion and science is related to God's action in the world, wherein the material scientists' understanding of nature as a closed system denies possibility of God's intervention. This position is championed by scientists like Laplace. The view of God's intervention in the world is also vigorously denounced by an astute philosopher of empiricism, David Hume. These opinions of Laplace and Hume were inconsistent with the Newtonian world view, which attributed certain periodic interventions of the planetary orbit to God.[16] The Laplacian thesis on cosmic design, however, gives no room for any divine involvement or intervention in his scientific work unlike Newton's gap theory. Laplace's view therefore strengthens reductionism in science, which is rudely averse to the claims of most religions that there is immateriality behind nature.

Whereas some discussion of the conflict between religion and science centers on what we believe about scriptures, others center on ethical decision making. "Ethics or moral philosophy, asks basic questions about the good life, about what is better and worse, about whether there is any objective right and wrong, and how we know it if there is."[17] Several ethical issues have been raised on virtually every area of modern science. It is obvious that "we use knowledge gained by science to help us (ourselves), but we are also subject to the demands of science. Modern technologies provide us with many goods and opportunities, yet in giving us more choice they also present us with difficult ethical questions."[18] For example, Parkinson's patients are treated using medications leading to brain damage;[19] the technologies to control conception; the treatment of genetic diseases by methods of turning off[20] replacement genes or changing malfunctioning genes; nanotechnologies; creation of 'techno sapiens'[21] and issues in biotechnologies that include advanced creation of cloned man[22] are serious issues where religious beliefs find modern technology disturbing. Barns hence opined, "the new biotechnology also represented an epochal challenge to Christian theology, of the same magnitude as the holocaust, not in the sense that biotechnology was a great evil, but in challenging many of the fundamental assumption[s] of a Christian understanding of' the human person and our place in nature."[23]

Barns therefore argued that the epistemological assumptions for religious thought were the main focus of open debates on techno science and will deepen the conflict between science and religion in relation to ethics. He stated

> The first is that an enduring instrumental positivistic conception of science. Despite the powerful arguments of social constructions about

> the cultural and political embeddedness of scientific research and technological innovation, the dominant discourse of techno science is still largely positivistic. The main-stream practices of science continue to tacitly assume a split between facts and values and the view that scientific investigation in itself is a value neutral exercise: even though it may be conceived that practitioners have a 'social responsibility' to consider wider social, legal, and ethical implications. The second assumption is of the basic "secularity" of the reality in which we are immersed. To be sure, it may be that the view that "religion" is going through a process of irreversible decline is now being seriously questioned by sociologists of religion. It may recognize that "religion" as a sociological phenomenon, both institutionalized and informal is alive and well. Nonetheless, the dominant epistemological framework of late modern societies (including that of revisionist sociologist of religion), is one in which reality is ultimately understood in secular and scientific terms.[24]

It is clear from this standpoint that the current friction between religion and science is gravitating toward the survival of religion in the face of secularism where science thrives. On the other hand science has found religion and its ethical garb as its main challenge from the time of Galileo until the present. It is clear from this background that the conflict of these two spheres is a significant problem in the development of a sustainable environment for human and economical development.

Man, however, is at the center in the religion and science conflict. In theological reflections, man has been seen as the crown of the entire creation. "Religion involves a relationship between man and that who is regarded as holy or divine . . . [and] man is a spiritual being who stands out from and above the rest of nature by his self-awareness and his capacity for self transcendence."[25] Man from religious perspectives will not just be concerned about life as it is, but his spiritual nature always pursues something higher. Thomas citing Vander Leeuw says

> When we approach it from the human side, 'from below—upward,' we observed that man never simply accepts the life that is given him but seeks to enhance its value [scientific knowledge] although he cannot find his ultimate meaning of his life in this way. His goal is salvation, a new and higher life. But in any case religion is always directed toward salvation, never towards life as it is given; and in this with respect to all religion, with no exception, is the religion of deliverance.[26]

The main essence of religion is therefore seen as helping man to attain deliverance from the mundane to supra mundane—the Holy—a *mysterium tremendum* before which the religious man feels awe or dread because of God's majesty.[27] Man's experience with the sacred demands a response of moral living from man. "Religion and Morality, at least in their highest forms have a common root or origin in human nature."[28] Therefore, there is a twin relationship between religion and ethics in the human life for him to be able to fulfill the divine purpose. Thomas argued

> The primary concern of religion is with the establishment and maintenance of man's relationship to the Holy or Divine and with the transformation of his life which results from that relationship. On the other hand, morality in its purest form arises from the pressure upon Man's consciousness of objective values and unconditional imperatives, with the ethics that believes to be, grounded in God, the transcendent reality. Its primary concern is with man's realization of the values and fulfillment of the obligations by which he is confronted. But the common aim of religion and morality is fullness of life.[29]

The interlocking relationship of ethical religion with concerns for man's fullness of life is not without the basic understanding of man's life after death. Virtually, all religions of the world are concerned about the afterlife. The idea of an afterlife in the civilization of ancient Greece was found in the Platonic conception of immortality.[30] The afterlife in the theistic campaign included the issue of the "consummation of the whole history of the world towards which all God's redemptive acts in history are directed."[31] Other religions such as Islam vary in their perspectives of eschatology. However, the fact of the belief in the consummation of the world is a concern for both Christianity and Islam. The idea of ethical living on the grounds of the afterlife are shared by most religions in the world, although some variances occur. However insignificant the eschatological differences, it also contributes to the conflict between religion and science. Haught observed that

> The synthesis of science with tragic mythology, a blend that we may call cosmic pessimism, still poses a serious challenge to all eschatological or hopes -filled interpretation of reality, and those whose faith forbids acquiescence in the seemingly more realistic stoicism onto which so much modern science has been grafted.[32]

Material science hardly bothers with the implication of the scientific creation of robots, cloned humans and other issues in the understanding of religious men. The large scale of destructive explosives made possible by technology and nuclear weapons along with other biological weapons bothers religious man in his eschatological understanding. It is however important to note that it is difficult to halt scientific inquiries especially with the vested interest of capitalists because

> In order to fulfill their objective of acquiring more and more capital, capitalists took a greater interest in the laws of science which could be

> harnessed in the form, of machinery to work and make profit on their behalf . . . today the quest for the profits comes into sharp conflict with people's demands that their material and social needs should be fulfilled. The capitalist bourgeois class is no longer capable of guiding the uninhibited development of science and technology—again because the objectives now clash with the profit motive.[33]

This conflict as it stems from the capitalist interest in making profits does not stop at separating the realm of science alone but also raises environmental concerns that propelled several religious concerns from ecologists and eco-feminists. The vested interest of capitalists is a major factor in the widening of the conflict between religion and science.

AFRICA IN THE CONFLICT BETWEEN RELIGION AND SCIENCE

The conflict between religion and science that pervaded the modern world is a dual concern in Africa. First, the mixed religiosity of an average African owing to imported religions widen the context of religions engaged in conflict with modern science and technological innovations in the minds of Africans. Second, because African ethical theistic religions with a clear eschatological understanding exist, the innovations of modern science, which tampers with the hubris of that religion and its cosmological inclination of oneness of spirituality and nature, is no less a conflict. Africa found itself in the religion and science conflict thanks to Western civilization as

> The eighteenth century saw itself as the Age of Reason in which the ideal of rationality manifest in science would permeate all human activities. . . . Its most articulate and extreme spokesmen were the French intellectuals of the mid-century, but its spirit permeated Germany, England, and the American colonies and influenced the subsequent climate of thought throughout the modern world. From among the many ideas of the Enlightment . . . God became a debatable hypothesis.[34]

Before the advent of Western missionaries and subsequent colonization, "the Africans are rather guided by their religions and not by laid down rules of their ancestors or heroes because "in all things, they are religious."[35] Africans had religions before the advent of both Christianity and Islam. Africans' understanding of foreign religions cannot be completely divorced from the indigenous religions that shaped his life and culture. Healey and Sybertz (1997) observed that

> The encounter of African culture with Christianity brings a newness, freshness, and originality, a difference like a spice that brings a new

> taste of food. In one way it is the same Food (core Christianity), but it taste[s] yet different. Deep African values, such as community, hospitality, the "living dead," patient endurance in adversity, and holistic living bring something, new and truly enrich world Christianity and the world church.[36]

African people also had cultural ethics before the advent of foreign religions. The belief in God as taught by their religions determined their entire lives. A typical African, for instance, would avoid any debate on the reality of God's existence because according to his or her religion, "it is the dangerous beginning of irreverent inquisitiveness."[37] This is similar to the awesomeness of God that humbles man as pointed out by Rudolf Otto in his thesis "The Idea of the Holy."[38] The attribute of God as a judge is significant to Africans and it enforces them to abide by the moral dictates of their indigenous religions. God is "All-wise, All–knowing (and) because He is all–seeing, his judgment is impartial. We often hear the expression '*Olorun mu un*.' He is under the lashes of *Olodumare* (God) when a misfortune befalls a person who is known to be a moral offender,"[39] just as Christianity established the fact that God judges the wicked like in the case of Herod (Acts 12:9–24). It is also noted among the Ashanti that "the ancestral beliefs and the practices did not die out with the turbulent events that came with Christianity and westernization."[40] Typical Africans believed in the world of the ancestors. There is a common saying among the Yoruba that *aye lo ja, orun ni ile* (The world is a market place, heaven is our home, just as understood in 2 Cor. 5:1). This shows that a typical African believes in heaven and an afterlife. However, the eschatological understanding among Africans varies considerably with Christianity.

Foreign religions came with their own culture, science and technology that were imposed on African cultures, cosmologies and technologies. Africans from the onset believed in "the twin-ness and cooperation derived from a great unity between the cosmos and earth. Ancient and traditional African people of color viewed humanity and the cosmos as one."[41] This worldview of Africans that accepted a dualistic approach to science was affected by the imposition of modern science. "African Sacred Sciences are like water that fill the dry desert of the spiritual void, created in the soul of Africans and other people affected by western minds."[42] "When it comes to the question of manufacturing in Africa before the time of the white man . . . Most African societies fulfill their own needs for a wide range of articles of domestic use, as well as for farming tools and weapons."[43] The Western technology imposed on Africa rubbished her homegrown technology that would have developed within the context of African ethical theistic religions.

It is worthwhile to note that Europeans did not teach Africans their science and technology, but rather they used it as a decisive weapon for colonization. They were also aware of the liberation that technology would bring to Africa; therefore, they were decisive, and consequently, adverse to teaching Africans new technology.[44] It was observed however that Africa was prevented from accessing her own technology by the instrumentality of capitalism:

> For instance, in the German Democratic Republic, the youth established a young innovators' fair in 1958. Calling upon the intellectual creativity of socialist youth, so that within ten years over two thousands new inventions were presented at that fair. The connection between Africa and Europe from the fifteenth century onwards served to block the spirit of technological innovation both directly and indirectly.[45]

The development of material science however was not strengthened by African religions similar to other religions of the world. Rodney observed that "African ancestral religions were no better or worse than other religions as such because of its non-scientific and metaphysical way of the universe which often conflict with materialist's outlook of societal development in terms of economy."[46] This is the very reason why Europeans limited religious dominance on African societies to give birth to secularism with the aim to speed up modern science's popularity, especially materialistic modern science, which has the potential to make gains of wealth and power for capitalists. Consequently, Western religion and especially African religions, because of its overall dominance on the Africans, were put into conflict with modern science.

As a recent study suggests, a major factor in the decline of science in the Islamic world was unconnected with the tight control of Islamic religious authorities over higher education.[47] Europeans governing Africa brought an end to religion's precedence in politics, geography and medicine; hence, secularism was motivated by Europeans' interest for quick development and the acceptability of material science created a world apart from religion. This of course spelled confusion to the then colonized Africa, which gave room to the eventual catalytic development of capitalism.[48] As a result, the gap between religion and science in Africa in regards to the understanding of the material world apart from the spiritual world became widened. Thus, belief in prayer and the intervention of the ancestors and various gods were substituted by innovations designed to control the effects of weather and the environment.[49] Consequently, Africans whose technology and spirituality were intertwined found themselves striving to purchase Western technology and products even at the risk of their land and worse still at the price of their people sold to slave traders.

IMPACT ON HUMAN AND ECONOMIC DEVELOPMENT

The estranged relationship between religion and science deepened via capitalist colonization and this negatively affected both the human and economical development of Africans: as sustainable development "is as much about economic and social (human) as it is about environmental protection."[50] The European colonial education system unfortunately did not teach Africans the development of science and technology. "It was not an educational system designed to give young people confidence and pride as members of African societies, but one which sought to install a sense of difference towards all that was European and Capitalist."[51]

Missionary education taught children how to read the Bible or the Quran and to be a priest. Therefore, science and technology to an African was a 'white magic' desired but not attainable. Worse still Christianity and Islam imposed their religion and technology at a period when the rising modern science of the eighteenth century and religion were in conflict. The associated problem of capitalist colonization such as poor education and deprivation of the development of home grown science and technology was glaring. This bred more conflict between religion and science.

Unfortunately, the conflict was rather staged more on African minds than anywhere else. This automatically shut African minds to the development of science and yet they craved voraciously for the consumption of Western technology. However, African hearts were opened to religion in the development of life and culture. Whereas this is good, it is also essential for Africans to build human and economic resources through an equal development of science and religion. The rather unfortunate scenario in modern civilization is the tendency for African people to fall hook, line and sinker for the bait of technological development at the risk of religion, which has shaped the values and culture of Africans. This is the perceived lopsided conflict between religion and science at the present moment in African development.

There is need of caution for policy makers not to awaken the rude strength of religious fanaticism in large-scale devastation such as the world experienced on September 11th, 2001, where a group of people recognized to be religious fanatics destroyed one of the world's largest economic and technological centers: religious bigotry and fanaticism rode on the horses of modern science to erupt havoc. Because fanaticism can be a response to secularism, which paved the way for material science, and knowing full well that its tendencies can be reinforced by conflict between religion and science, particularly among Africans, because of the very strong influence religion can have over them, hence, there is need for a balanced understanding of the two spheres as Africans rise to technological development to improve their environment and economy.

The economy of developed nations was made possible by the application of science and technology to production. "At the beginning of the Colonial period, science and technology as applied to production already had a firm base inside Europe—a situation which was connected to overseas trade."[52] There is, therefore, a necessity for Africans to be supported and not exploited in the development of their home grown science and technology in the context of their religious and cultural heritage to bring about a sustainable and not lopsided world economy as currently witnessed today. The human and material resources must therefore be consciously developed to attain a desired peaceful and economically viable ecosystem.

STEPS TOWARD RECONCILIATION BETWEEN RELIGION AND SCIENCE

The discussion in this chapter clearly shows major areas of conflict between religion and science. There are, however, indications for possible steps to make religion and science allies to allow a sustainable environment for both human and economical development, especially in Africa. There are roles for both religious and scientific experts to play in engineering a sustainable economic and human environment. Three key steps are discussed below.

First, there must be a more convergent effort toward a crystallization of African contextual theology. It is very important to establish the theological framework for ethics and values that are indigenously African in which an alliance between religion and science can be formed: applying the principle of propinquity and proportion because there is a lot of similarity found among African indigenous religions and cultures. "It is not a matter of talking the traditional customs of African culture and melting best ones pot into Christianity. It is not a matter of African culture tactics being mediated through western culture and through patterns."[53]

The contextualization of any foreign religion must be with a clear understanding of the African people without prejudice to the salvation message significant to Africans. This will go a long way in building African science and technology that will be in harmony with society just as is happening in the fields of Art and Humanities. Healey and Sybertz (citing Elochukwu Uzukwu) write

> African artists are beginning to speak their communities' faith through images. The painting of Mary and the Child Jesus dominating the sanctuary in the cathedral of Yaounde, the carvings and painting in the cathedral of Kinshasa, the wood-carved Stations of the Cross in the visitation sisters' chapel in Loango, Congo, the varieties in vestment design found all over Africa, testify to a bourgeoning of African liturgical art.[54]

Contemporary churches have some of these images carved by African artists. The music and home video industries are beginning to imbibe some of the values of African culture. This similar development can be made possible in science and technology. The development of African contextual theology will help to correct such ideas that a 'white man' is in all things way better than the 'Blackman.'

Secondly, there must be encouragement for the development of home grown science and technology. Even where the ideas are borrowed from the Western world, such should be properly substituted or integrated into African culture and the environment. It is only such decolonized science and technology that can bring us to a proper perspective of what we can do as a people in building a sustainable environment that will encourage progressive human and economic development. The 'African magic' must be developed to help Africans be less dependent on the 'white magic,' which has deeply underdeveloped black people and their continent.

Thirdly, the conflict between science and religion must be cursorily studied by Africans. It is important for Africans to participate in the present global perspective of the debate between religion and science. At this present level of globalization, which makes the world a global village, it is imperative to wisely follow the global trend as it affects Africans and simultaneously keep working at the local level to develop Africa. The present condition of the debate between religion and science affects every nation, especially the developing world. Some scientific studies have found that the economics of poor countries will suffer most from greenhouse effects.

The influence of science and technology exported to the African continent demands that we engage in this study. Because Africans today have come to the awareness of modern Western civilization, being consumers of Western scientific theories and technology, they cannot but be concerned with the Western world on issues related to evolution theory, which reduces man's ancestry to apes; reductionism of the human soul, cloning, artificial insemination that challenged the traditional biology of human existence, creation of robots, cyborgs and environmental challenges such as the greenhouse effect and other disasters. The contribution of Africans in this field of global concern will significantly enable a critical examination of the inherent conflicts that may manifest between religion and science as the investment of resources in the development of African people and their economy is increasingly encouraged.

The model of the conflict between religion and science that was seen in the area of the mechanistic worldview of Laplace and Darwinian evolution remain historically undeniable. The bottom line, which is both ethical and eschatological for religion, for science is a case of conquering nature and creating a better conducive environment for man through technology. It is also undeniable that modern science brought wealth and power to the developed world and contributed to the underdevelopment of developing nations. The conflict between science and religion in Africa is therefore an undeniable historical truth and more so, a pathetic one because it actually stemmed from colonization and its interest to create wealth and power, and the missionary theological orientation, which was on one hand an opposition to most cultural African beliefs and on the other hand at loggerheads with the rise of modern science. Should Africans have been allowed to govern themselves during that period, strong indications suggest that their pre-colonial scientific worldview would have been beneficial to the development of modern science in the fifteenth century. It is therefore imperative to balance the religious and scientific understanding of the African mind to cause a catalytic development of both the people and the economy. A reconciliation of African religions, culture and science is imperative in an age of globalization to prevent an avalanche of imported, value-free or unacceptable ethically coded science and technology.

NOTES

1. John F. Haught, *Science and Religion: From Conflict to Conversation* (New York: Paulist Press, 1995), 3.
2. Ian. G. Barbour, *Religion and Science: Historical and Contemporary Issues* (New York: SCM, 1998), 24–25.
3. Walter Rodney, *How Europe underdeveloped Africa* (Washington, D.C.: Howard University Press, 1982), 35.
4. Barbour, xiii–xv.
5. Some Of Einstein's Writings On Science and Religion, "Science and Religion," 1959, http://www.sacred-texts.com/aor/eistein/einsci.html. Accessed 02/15/09.
6. John H. Brooke, *Science and Religion: Some Historical Perspectives* (Cambridge: Cambridge University Press, 1991), 2.
7. Ibid., 2.
8. Rodney, *How Europe Underdeveloped Africa,* 4.
9. E. Bolaji Idowu, *Olodumare: God in Yoruba Belief* (Nigeria: Longman, 1996), 22.
10. Joseph I. Omoregbe, *A Philosophical Look At Religion* (Lagos, Nigeria: Joja Press Limited), 19.
11. Eugene C. Scott, "The Science, Religion and Evolution Workshop Reader Science and Religion Course Program," (Berkeley, Calif.: CTNS, 2001), 363.
12. *Stanford Encyclopedia of Philosophy,* "Religion and Science," 2007, http://www.sep.org/religion/science/mhttml. Accessed 03/10/09.
13. Ibid.
14. Eugene C. Scott, "The Science and Religion Movement: An Opportunity for Improved Public Understanding of Science?," *Skeptical Inquirer* (1999): 31, in "Workshop Reader, Science and Religion Course Program" (Berkeley, Calif.: CTNS, 2001).
15. *Stanford Encyclopedia of Philosophy Religion and Science,* http://plato.stanford.edu/entries/religion-science/. Accessed 03/12/09.
16. Ibid.
17. Barbara Mackinnon, *Ethics: Theory and Contemporary Issues* (New York: Wadsworth Publishing Company, 2003).
18. Ibid., 391.
19. Ibid., 390.
20. Ibid.
21. Techno Sapiens is taken from the Ninth European Conference on Science and Theology (EC ST IX) in Nijmegen, Netherlands (2004).
22. Mackinnon, 392–393.
23. Ian Barns, "Theological Implication of New technologies," *Theology and Science* 3, no. 2 (2005): 180.
24. Ibid., 181.
25. George F. Thomas, *Philosophy and Religious Belief* (New York: Charles Scribner's Sons, 1970), 49.

26. Ibid., 49.
27. Ibid., 52.
28. Ibid., 57.
29. Ibid.
30. William L. Rowe, *Philosophy of Religion: An Introduction,* 4th ed. (New York: Wadsworth, 2006), 147.
31. John F. Haught, "Evolution, Tragedy and Hope," in *Science and Theology: The New Consonance,* ed. T. Peters (Colorado: Westview Press, 1998), 234.
32. Ibid., 230.
33. Rodney, 10.
34. Barbour, 34.
35. Idowu, 22.
36. Joseph Healey and Donald Sybertz, *Towards an African Narrative Theology* (Limuru-Kenya: Pauline Publications Africa, 1996), 20.
37. Idowu, 17.
38. Thomas, 37.
39. Idowu, 39.
40. J.O. Omosade Awolalu and P. Adelumo Dopamu, *West African traditional Religion* (Ibadan, Nigeria: Onibonoje Press, 1979), 10.
41. Oba T'Shaka, *Return to the African Mother Principle of Male Female Equality, Vol. 1* (Oakland, Calif.: Pan African Publishers and Distributors, 1995).
42. Ibid., 90.
43. Rodney, 40.
44. Ibid., 105.
45. Ibid.
46. Ibid., 9.
47. Barbour, 27.
48. Rodney, 35–36.
49. Ibid.
50. Ibid.
51. Ibid., 240.
52. Ibid.
53. Healey and Sybertz, 19.
54. Ibid., 23.

Part II

Oil and Politics

5 Oil Multinationals

'Environmental Genocide' and Socioeconomic Development in Nigeria's Niger Delta

Donald O. Omagu

INTRODUCTION

The Niger Delta was the locus of commerce in palm oil and general merchandise in pre-colonial and colonial times. Currently, the Delta has great strategic value because of its oil and gas reserves. Crude oil accounts for 95 percent of Nigeria's foreign exchange earnings and about 25 percent of the gross domestic product. Besides the great fuel wealth, the Delta also possesses arable land, forests, rivers and coastal waters teeming with fish and sundry aqua-cultural resources. The Delta is one of the most ecologically rich and diverse regions in the world. It covers an area of about 112,110 square kilometers situated in the southern part of Nigeria and bordered by the Atlantic Ocean to the south and Cameroon to the east; it includes about 12 percent of the country's area.[1] The region is home for the Ijaws, Okirikas, Kalabaris, Efiks, Ibibio's, Urhobos, Igbo, Edo, Ukwuanis and Itsekiris and so forth. Prior to the advent of the British colonial government, the different ethnic groups enjoyed unbridled monopoly over the waters in their areas. They used the rivers and the sea for their economic advancement.

Since 1956 when oil was struck in Oloibiri, the Niger Delta and its resources have sustained Nigeria's economy. Since 1970, Nigeria has earned at least $300 billion from energy development. It took in $45 billion in 2005 alone. With about 40 billion barrels of proven oil reserves, Nigeria currently produces about 2.4 million barrels of oil per day. It is the fifth largest source of oil for the United States.[2] Yet, it still has untapped resources of over 30 billion barrels of oil and gas reserves of about 16 trillion cubic feet. Besides its natural resources, the Niger Delta is also blessed with a natural environment that is conducive for tourism and agricultural production and other wealth-producing activities that, if properly harnessed, could bring fortunes not only to the region and its people but also to the entire country. The pristine serenity of the Delta environment has always attracted people from within and outside Nigeria. The region was quickly located by the first white men who came to these parts because of its easy accessibility by water.[3]

The Delta's environment can be broken down into four ecological zones: coastal barrier island, mangrove swamp forests, freshwater swamps and lowland forests. This incredibly well-endowed ecosystem that contains one of the highest concentrations of biodiversity on the planet, in addition to supporting abundant flora and fauna, arable terrain that can sustain a wide variety of crops, lumber or agricultural trees and more species of freshwater fish than any ecosystem in West Africa. Yet, in spite of its considerable natural resources, the area is one of the poorest and most underdeveloped parts of the country; 70 percent of the inhabitants still live a rural, subsistent existence characterized by total absence of such basic facilities as electricity, pipe-borne water, hospitals, proper housing and motorable roads. They are weighed down by debilitating poverty, malnutrition and disease. Whereas decades of corruption and mismanagement in the echelons of power have plunged the country's gross national product per capita to an all-time low of $280, annual incomes in the Niger Delta are still far below the national average. The area also has one of the highest population densities in the world, and annual population growth is currently estimated at 3 percent. Rapid population growth increasingly exerts pressure on cultivable land, a good part of which is in any case prone to flooding almost all year. The population of Port Harcourt and the other major towns is literally exploding. The ensuing scenario—urbanization without the economic growth that would ordinarily generate more jobs—has resulted in the human ecologist's ultimate nightmare of a growing population that, in a bid to survive, is destroying the very ecosystem that should guarantee its survival.[4]

In addition, the carelessness of the oil industry has precipitated a situation where the region could most likely experience the loss of its inhabitable terrain, the poisoning of the water and the destruction of the vegetation and agricultural land by oil spills that occur during petroleum operations. This dismal picture of environmental pillage in the Niger Delta, despite its enormous endowment in oil reserves (that should bring promise rather than penury), is the responsibility of five multinational firms that care naught for the consequences of their actions: Royal Dutch Shell (Britain), Total (France), Azienda Generale Italiana Petroli (AGIP; Italy), Exxon Mobil and Chevron (United States). The behavior of those corporations, especially the lack of environmental oversight, is responsible for the Delta's destruction.

In the course of exploring for oil in the Niger Delta, Shell, contrary to what the various public relations agencies and consultants in its employ have been striving to sell to the International community, has not only radically disrupted the ecological balance of the area, but through negligence and cynical difference has orchestrated a vicious ecological war—a war whose victims are a hapless people and the land on which they have lived and thrived for centuries. The report submitted to the world conference on indigenous people on Environment and Development during the Rio Earth Summit in June 1992 by the Kings, Chiefs and community leaders of the

Niger Delta told the harrowing story of the still ongoing ecological war waged by shell in the Delta.[5]

> Apart from air pollution from the oil industryís emissions and flares day and night producing poisonous gases that are silently and systematically wiping out vulnerable airborne biota and otherwise endangering the life of plants, game, and man himself, we have widespread water pollution and soil and land pollution that respectively result in the death of most aquatic eggs and juvenile stages of life of finfish and shellfish and sensible animals (like oysters) on the one hand, whilst, on the other hand, agricultural lands contaminated with oil spills become dangerous for farming, even where they continue to produce any significant yields.[6]

Similarly, Robert Bullard observes that the oil-producing states within federated Nigeria have benefited the least from oil wealth. Devastated by the ecological costs of oil spillage and the highest gas flaring rates in the world, the Niger Delta is a political tinderbox. The militancy of a generation of restive youth, the deep political frustrations among oil-producing communities and pre-electoral thuggery all prosper in the rich soil of political marginalization.[7]

This chapter examines decades of the heinous history of unmanaged oil exploration with the appurtenant hazardous spills and gas flares by oil firms and their environmentally hostile activities in the Niger Delta, including the damage to the health and economic viability of the regionís people. Lamentably, hazardous spills and gas flares have not only degraded the environment and destroyed farming and fishing, but the major occupations of these predominantly river communities still lacks the elementary infrastructure and basic communal necessities: electricity, roads, schools, hospitals, portable drinking water and so on. The long years of neglect and deprivation, coupled with the insensitivity of some previous governments and oil companies, as well as the failure of previous development intervention agencies, had created, by the late 1990s, a volatile atmosphere characterized by protest, agitation and communal conflict.

OIL EXPLORATION AND ENVIRONMENTAL POLLUTION

Since 1958, the region's natural beauty and wealth has become more of a curse than a blessing to its inhabitants owing to ecological disaster caused by the activities of Shell and other multinational oil companies, including AGIP and Mobil.[8] Assessing the environmental destruction is not easy because few companies want to report their emissions and oil leaks especially when they are high; nor do they want an exposé written about the damage they inflict and have it published worldwide. A great deal of the damage occurring in Nigeria is known because community groups took grievances to the world and environmental activists from all over joined their resistance movement. In Iraq and Saudi Arabia, the extent of the damage is not clear. Partially this is because much of the Middle East consists of uninhabited desert and the direct impact on people is minimal. For this reason, assessing the environmental impact of an oil industry can be difficult, but it is by no means impossible.[9] A perfect example of a successful attempt at such an assessment is the Berger Commission enquiry before the Mackenzie Valley pipeline was built in the Northwest and Yukon territories of Canada. In that case, Ottawa listened to the people involved! Hence, the pipeline was designed and constructed largely in response to the proposals of the people rather than oil barons.[10]

The multinational oil companies, mainly Shell, Chevron/Texaco and Elf, through their activities have systematically transformed the Niger Delta into an ecological disaster zone. Oil spills result from corrosion of oil pipes, poor maintenance of infrastructure, leaks and human error and at times as a consequence of vandalism, theft of oil or storage.[11] Michael Fleshman, of the New York-based Africa Fund, describes what he saw at the site of one oil spill:

> The impact of the spill on the community has been devastating, as the oil has poisoned their water supply and fishing ponds, and is steadily killing the raffia palms that are the communityís economic mainstay. Lacking any other alternative, the people of the village have been forced to drink polluted water for over a year, and the community leaders told us that many people had become ill in recent months and that some had died. The sight that greeted us when we finally arrived at the spill was horrendous. A thick brownish film of crude oil stained the entire area, collecting in clumps along the shoreline and covering the surface of the still water. The humid air was thick with oil fumes.[12]

The common response of the oil companies to such spills, however, has been to blame the villagers for sabotage. Sabotage and theft through oil siphoning has become a major issue in the Niger Delta as well, contributing to further environmental degradation. Damaged lines may go unnoticed for days and repairs of the damaged pipes takes even longer. Oil siphoning has become a big business, with the stolen oil quickly making its way onto the black market.[13] A consultantsí report for Shell in 2003 estimated that a staggering 275,000 to 685,000 barrels per day of crude were being stolen by oil thieves in the Delta. At $50 a barrel, this would mean that over $5 billion a year of crude oil is stolen, much more than the Angolan rebel leader Jonas Savimbi earned from his diamonds and used to buy the missiles, armor and artillery that kept the oil-rich Angolan government and his MIG fighter-bombers at bay for years. Bunkering fuels huge arms flows, amid fierce turf battles for access to this trade. It is no surprise that the U.S. Defense Department is now taking serious interest in the Niger Delta.[14]

Every major oil company has experienced serious accidents on rigs everywhere. In the delicate ecosystem of the Niger Delta, these accidents cause grave damage to the environment. A gas explosion at Oleh oil field in July 1995 resulted from a faulty pipe. The ensuing fire engulfed the adjoining bush and burned down valuable trees. Another oil spill occurred in June 1999 when a pipeline buried beneath the deep waters of the Lake Lion Onome burst, spewing oil. Far flung communities in the path of the tidal waves, including Olomoro and Emede, were badly affected with food crops like cassava, yam and coco-yam destroyed. Thousands of fishponds as well as natural and artificial lakes were contaminated by crude oil. Fishing nets and baskets were ruined.[15] It is estimated that nearly six hundred people have died in oil related accidents worldwide since 1965. Union Oil's 1969 Platform Alpha blowout may have been the most serious offshore accident in its effect on U.S. politics, but it hardly ranks among the most serious offshore disasters. Twenty years after Platform Alpha's blowout, 167 people were killed when Occidentals Pipe Alpha platform exploded off the Coast of Aberdeen, Scotland, in 1992. Oil rigs are alighting all along this stretch of Africa's western coastline like giant metal mosquitoes, standing on the skin of the earth on spindly legs and drilling down with steel proboscises to suck out the fluid that is the lifeblood of the world economy. Like the biting insects, the rigs can cause irritation around the site of extraction, disrupting local communities or polluting farmland. But it is this resource curse—the stealthier, time delayed payload that accompanies the extraction, just like the malaria that real mosquitoes transmit—that is the real threat.

Oil spillage has a major impact on the ecosystem. Immense tracts of the mangrove forests, which are especially susceptible to oil, have been destroyed. An estimated 5–10% of Nigerian mangrove ecosystems have been wiped out either by settlement or oil. The proportion of crude oil lost to the environment when spills occur is quite disturbing. A Department of Petroleum Resources report in Nigeria states that over 95% of the volume of oil spilled in the region is not recovered. This spells serious health hazards for the communities in the area affected. It is *prima facie* evidence that no elaborate and concrete efforts are made to protect the ecosystem from environmental pollution.

The rain forest, which previously occupied some 7,400 km of land, has disappeared as well. Spills in populated areas often spread much further by taking out crops and aquacultures through contamination of the groundwater and soil. The consumption of dissolved oxygen by bacteria feeding on the spilled hydrocarbons also contributes to the death of fish. In agricultural communities, often a year's supply of food can be destroyed by only a minor leak, debilitating the farmers and their families who depend on the land for their livelihood. Drinking water is also frequently contaminated, and a sheen of oil is visible in many local bodies of water. If drinking water is contaminated, even if no immediate health effects are apparent, the numerous hydrocarbons and chemicals present in oil are highly carcinogenic. Consequently, the people have traditionally depended on rain water for drinking and bathing. But the rain has been acidic for more than a generation because of gas flaring. People have since resorted to well water, which is already polluted by both oil spillage and acid rain. Respiratory problems, coughing up blood, skin rashes, tumors, gastrointestinal problems, different forms of cancer and malnourishment witnessed in many communities can be linked to environmental pollution. Many children have distended bellies and light hair, which are evidence of kwashiorkor, a protein deficiency syndrome.[16] There are parallels between the Niger Delta and the State of Louisiana. Louisiana's petroleum industries are extremely important to the state and the nation: for decades, these industries have made significant economic contributions to the state. The pollution burdens created by these industries, however, are immense. Corridor residents have to endure greater health risks and sometimes die from diseases they believe are related to chemical exposure. An even more sinister consequence of the growth of the corridor has been the loss of historic lands and communities.[17]

In assessing the crisis, the World Bank stated that over 2,300 cubic meters of crude oil in 300 separate incidents was spilled into the River (including Bayelsa and Delta) states between 1991 and 1994. Since 1996, the graph of oil spilled into the environment increased sharply with attendant costs to the environment, the economy and future generations. This was similar to Texaco's (now Chevron Texaco) twenty years (1971 to 1991) of drilling for oil in the Oriente (Ecuador's Amazon region), one of the most ecologically rich and diverse regions in the world. By all accounts, save those of Texaco, the drilling has left an enormous mess. The company, it has been widely alleged, spilled 16.8 million gallons of oil (one and half times that of the 1989 *Exxon Valdez* oil spill in Alaska) from ruptured pipelines, discharged 19 billion gallons of highly toxic water into waterways and the soil, conducted massive deforestation and left more than 600 toxic pits uncovered.[18]

The Federal Environmental Protection Agency Act enacted in 1988 was the first Nigerian statute to deal exclusively with the environment. The Act established the Federal Environmental Protection Agency (FEPA), which is responsible for the establishment of national environmental standards. Yet, FEPA officials have not been able to apply legal sanctions to compel oil companies to pay compensation for damage done to the environment or to restore polluted environments to an acceptable level.

Protected by the federal government of Nigeria, the oil companies operated in the delta for years with little environmental oversight. There was no federal environmental protection agency until 1988, and environmental impact assessments were not mandated until 1992. Although FEPA was set up in 1989 to monitor environmental abuse, in the case of the oil industry in Nigeria, however, FEPA is an idle threat. In spite of the fact that the activities of the oil operators take place in territories belonging to states, they have no authority by law to penalize polluters of the environment. The reason is simple: oil and mining matters are the preserve of the federal government as indicated in the Exclusive Legislative List of the Constitution.[19]

Ironically, if there is any environmental damage in Britain, the United States, France or the Netherlands, for instance, that results from the activities of the oil and gas companies whether they are directly or indirectly linked to such pollution, they must pay compensation in their own country. But when there is environmental pollution in the Niger Delta, compensation is rarely paid to the affected communities. Instead, the communities are blamed for causing such pollution through the vandalism of oil installations. The oil companies are bound by law to clean up the mess they create in the course of the operations. In Nigeria, the law, when it exists, is either full of loopholes or observed in the breach.[20]

Worst still is the attitude of multinational corporations toward mop up exercises and environmental impact assessments. These studies are rushed by academic contractors and focus on microbiology, chemistry and pollution; some include wildlife and fish but are done by people who are unfamiliar with the Delta fauna. In some instances contracts are awarded to influential members of the community who have little or no idea about how to clean up the mess.

The exploration and exploitation of oil go along with specific environmental hazards. The drilling activities are land based, including the swampy areas. They cut the well formations, cushion them and use drilling muds: cement slurry and other acid stimulation are also used. The wastes from these sources are harmful to the flora and fauna of the environment, including human beings. They are poorly managed and in most cases not treated. Besides, the area is regarded as a minefield: the companies dump these untreated harmful wastes either directly on the land or into the swamp or dig pits into which they are disposed, pits that are not cased at the bottom and top and therefore causes overflow and spill into other areas mostly during rainy seasons. Of course, a cased and cemented well still needs a path through which petroleum can flow from the identified formation. Because most of the chemicals used by the contracted oil servicing companies are water based, the depth and speed of pollution are so fast in the swampy areas that within an average of twenty-four hours over five communities have their marshy areas polluted.[21]

Regrettably, human lives have been lost because of the unsafe corporate practices by oil multinationals. For instance, owing to the proximity of Shell facilities to human habitation, it is not unusual for children to fall into Shell oil wells. This was the fate of James Egbor, a secondary school student from Oleh, who in 1996 fell into one of the pits and died instantly. On their part, Shell and other oil companies have always said that they are not responsible for the majority of the calamities that befall the Niger Delta people. They put the blame of most spills on saboteurs who break oil pipes for 'quick business' or who register their displeasure against Shell and the government. They also claim that their company's investments have brought wealth to the country and therefore they should not be regarded as exploiters.[22]

Pollution is caused by the emission of pollutants from refineries, gas flaring at production wells and refineries and the use of petroleum fuel.[23] Gas flaring on a large scale is an extremely wasteful and environmentally harmful practice that is now universally discouraged as it is a major contributor of greenhouse gas.[24] Nigeria flares more natural gas associated with oil extraction than any other country in the world. Estimates are notoriously unreliable, but roughly 2.5 billion cubic feet of gas associated with crude oil is wasted in this way everyday. This is equal to 40% of all Africa's natural gas consumption in 2001, whereas the annual financial loss to Nigeria is about U.S. $2.5 billion. The flares have contributed more greenhouse gases than all of sub-Saharan Africa combined. And the flares contain a cocktail of toxins that affect the health and livelihood of local communities, exposing Niger Delta residents to an increased risk of premature deaths, child respiratory illnesses, asthma and cancer.[25]

In fact, in Western Europe, 99 percent of associated gas is used or reinjected into the ground. Gas flaring is generally discouraged and condemned by the international community as it contributes greatly to climate change, which ironically can display its most devastating effects in developing countries like Nigeria, and particularly in the semiarid Sahel regions of sub-Saharan Africa. The Niger Delta's low-lying plains are also quite vulnerable, as they lie only a few meters above sea level.

Gas flares can have potentially harmful effects on the health and livelihood of the communities in their vicinity, as they release a variety of poisonous chemicals: nitrogen dioxides; sulfur dioxides; volatile organic compounds like benzene, toluene, xylene and hydrogen sulfide; as well as carcinogens like benzapyrene and dioxin. Humans exposed to such substances can suffer from a variety of respiratory problems, which have been reported among many children in the Delta but have apparently gone uninvestigated. These chemicals can aggravate asthma, cause breathing difficulties and pain, as well as chronic bronchitis. Of particular note is that the chemical benzene, which is known to be emitted from gas flares in undocumented quantities, is well researched as being a causative agent for leukemia and other blood-related diseases. A study done by Climatic Justice estimates that exposure to benzene would result in eight new cases of cancer yearly in Bayelsa state alone.

Often gas flares are located close to local communities, and regularly lack adequate fencing or protection for villagers who may risk nearing the tremendous heat of the flare to carry on their daily activities. Many of these communities claim that nearby flares cause acid rain, which corrodes their homes and other local structures, many of which have metal roofing. However, whether or not the flares contribute to acid rain is debatable, as some independent studies have found that the sulfur dioxide and nitrous oxide content of most flares was insufficient to establish a link between flaring and acid rain. Other studies from U.S. Energy Information Administration (EIA) report that gas flaring is a "major contributor to air pollution and acid rain."[26]

Most of these flareís burn 24 hours a day and some have been doing so for over 40 years! Communities near these flares are deprived of even the comfort of nightís natural darkness.[27] Almost no vegetation can grow in the area directly surrounding the flare due to the tremendous heat it produces. The effects of gas flaring on the native health care system in the Niger Delta is evident in the great depletion of their medicinal herbal plants. Apart from the knowledge of the medicinal value of the plants among the people, the area is equally rich in medicinal herbs, both in the aquatic and upland areas. The forest serves as the natural pharmaceutical gardens from which the indigenous doctors collect most of their herbs. But these herbs have almost been depleted as a result of the heat generated from continuous flaring of gas by oil companies.[28]

The impact of oil and gas exploration on the Niger Delta environment is aptly described thus:

> Oil slicks cover the region. Blow-outs and leaks affect creeks, streams, and related traditional sources of livelihood, poisoning the water supply, destroying mangrove forests, eroding soil plots, and killing aquatic life. Hundreds of well-sites have flares, which come from the burning of associated gas. The flares heat up everything nearby and turn day into night, releasing 25 million tons of CO2 and 12 million tons of methane annually. Resulting sulfuric acid mists damage plants and forests. Flares pollute rain water, cause acid rain, and contribute to climate change.
>
> Many residents suffer from oil poisoning. Crude oil enters the body through skin absorption, ingestion of food and water, and inhalation of oil and dust particles. Oil poisoning also causes respiratory ailments. In addition, residents suffer from a plethora of water borne diseases such as malaria, dysentery, tuberculosis, typhoid, and cholera. Life expectancy is low and child morbidity rates are high.[29]

Nigeria has continued to pass legislation and policy pronouncements banning and/or discouraging flaring of oil-associated gas since 1969, but the political will to actually curb gas flaring is lacking. This raises the issue of the influence of multinational corporations over natural resource exploitation in developing countries. In this regard, Nigeria is not an exception. What is happening in the Niger Delta would never be allowed in any industrialized nation of the West. Oil companies cannot be allowed to become parallel governments in the West, with the power to determine what they will do and will not do, whether they will clean up oil spills, and whether they will pay compensation to victims of oil pollution. Yet they act with impunity in the Niger Delta.[30]

In 1969, the Nigerian government enacted legislation that made it a requirement for oil companies to set up facilities to use the associated gas within five years of the commencement of oil production. The violation of this legislation necessitated the enactment of the Associated Gas Re-injection Act in 1979, which charged oil companies to reinject associated gas and stop the unhealthy flaring within five years. The legislation was and still remains unenforced owing to the obvious laxities in policy making and implementation in Nigeria's energy sector.

In response to continued gas flaring by multinational corporations, the Nigerian government later resorted to a fairly insignificant penalty in the form of fines, which made it more economical for the companies to flare than to utilize or reinject gas. The fine for gas flaring was fixed at NO.5 (U.S. 0.038 cent) for every 1,000 ft3 of gas flared. It was later increased to N10 (U.S. 0.077 cent).This fine imposed on multinationals in Nigeria is a far cry from N1,300 (U.S. 10 dollars) penalties required for the same quantity of gas flared in developed countries, which has discouraged the companies from flaring gas in developed regions of the world. The continuous, rampant gas flaring in Nigeria is attributed to the ridiculously low fine for gas flaring, which has made it cheaper to pay the penalty than build facilities to collect and transport the gas for use as fuel and for electricity generation.[31] This explains why Chevron stated that whereas gas flaring would cost the company $1 million, the costs of switching from water to gas injection would cost $56 million.[32] This could be profitably compared to Royal Dutch Shellís behavior in Indonesia just prior to World War II.[33]

Considering the significant health implications, and global warming concerns over gas flaring, the Nigerian government proposed the year 2003 for the termination of gas flares. The oil-producing companies jointly attacked the policy as technically unfeasible. Consequently, the 2003 deadline was extended to 2008, which was confirmed again by the government and oil multinationals as unfeasible despite a court order, obtained by some affected communities, directing the oil companies to put a stop to gas flaring by December 31, 2008. In 2009 the Nigerian senate passed a bill to end gas flaring by the end of 2010 but flaring continues.

In realization of the effectiveness of economic instruments in enforcing elimination of gas flaring, a new fine of U.S. 3.50 dollars for every 1,000 ft3 of gas flared came into force in April 2008. The reduction of flared gas from 2 billion standard cubic feet (bscf) to 1.5 bscf per day, as at December 2008, may not be unconnected with the new fine. Oil companies are beginning to find it more economical to reinject than flare the associated gas. It is hoped that continual increases of the fine would force the multinationals to put an end to the unacceptable gas flaring.[34]

SOCIOECONOMIC DEVELOPMENT

Although the Niger Deltaís energy assets provide nearly three-quarters of government revenues and roughly 90% of foreign exchange earnings, they have not resulted in socioeconomic development. Unemployment hovers at about 90%. Many schools and hospitals lack staff, equipment and necessary supplies. Most riverine areas remain unconnected to electricity grids. Bayelsa State has only one multilane, paved road in the entire state, which leads only to the capital.[35]

Nigeria is one of the most egregious examples of an oil-driven economy out of control, but it is hardly on its own. Just like Ecuador, Nigeria, along with all of the major oil-producing states that make up the Organization of Petroleum Exporting Countries (OPEC)—Algeria, Iran, Iraq, Kuwait, Libya, Qatar, Saudi Arabia and the United Arab Emirates—are all dependent on oil revenue for economic survival.[36] Sadly, in spite of the Deltaís natural resource endowment, its immense potential for economic growth and sustainable development, the region has one of the most crushing poverty levels in the world. The poverty level in the Niger Delta is exacerbated by the high cost of living. In the urban areas of Rivers State, the cost of living index is the highest in Nigeria, only one-fifth of rural housing is considered physically sound, waterborne diseases are common, electrification is poor and so is the water supply and sanitation. Resources like oil and gas should be a blessing for countries that produce it. Norway and Britain seem to have done well from their oilfields, but in Nigeria the record is different.

Nowhere are failures so profound and visible than across the oilfields of the Niger Delta. For the vast majority, oil has brought only misery, violence and a dying ecosystem.[37] Even by the most liberal standard of development, the Niger Delta has not fared better in the hands of the oil multinationals. A recent United Nations report confirms among other things that

> The Niger Delta produces the oil wealth that accounts for the bulk of Nigeriaís earnings. Paradoxically, however, these vast revenues from an international industry have barely touched the Niger Deltaís own pervasive poverty. For most people, progress and hope, much less prosperity, remain out of reach. If unaddressed, these do not bode well for the future of Nigeria or an oil-hungry world.[38]

In quality of life, Nigeria rates below all other major oil nations, from Libya to Indonesia. Its annual per capita income of $1,400 is less than that of Senegal, which mainly exports fish and nuts. The World Bank categorizes Nigeria as a "fragile state," beset by risk of armed conflict, epidemic disease and failed governance.[39]

Instead of investing in genuine community development projects, oil companies apparently put their money into dividing communities and destroying effective organizing for human rights. Outraged by oil spills in Ogoniland, Saro-Wiwa organized a mass protest against the oil industry through "The Movement for the Survival of Ogoni People," which he founded. When the Ogoni launched a bill of rights in 1990, rebuking the oil companies for polluting their land and sowing divisions, the plight of the Delta began to find sympathetic ears in the West, where the public had been sensitized to the horror of oil's environmental impact by the Exxon Valdez oil spill disaster of 1989. Saro-Wiwa, like many Africans, understood Western sensibilities far better than Westerners understood Nigeria, and he exploited the environmentally sensitive mood. His book *On a Darkling Plain,* about the treatment of minority ethnic groups, added locally to the flammable mix. The system of revenue allocation and the insensitivity of the Nigerian elite, he wrote, "have turned the (Niger) Delta and its environs into an ecological disaster and dehumanized its inhabitants." Violence spread in Ogoniland, in the first strings of a new threat to the world's energy supplies.[40] Alarmed by Saro-Wiwa's popular support, Nigeria's military government arrested him and fifteen others and accused them of incitement to murder following the deaths of four Ogoni elders. The activists were denied due process and despite entreaties from world leaders, he and his comrades were executed in November 1995.

Oil companies have a long history of doing very little in the name of community development. It was only in the wake of Ken Saro-Wiwa's death that significant resources were deployed, but the rhetoric of corporate social responsibility obscures a much more pernicious set of practices. The policy of "cash payments"—used to pay for protection services from local unemployed youth, to buy off local opposition and to feed a vast shadow-world of illicit activity—have had the effect of generating enormously violent conflicts among and within youth groups. Some of these payments are made to militias who have "security companies" of their own. Young men who are given "standby payments" or hired as local security inevitably engage in struggles among themselves to capture or preserve the rents from competitors (other youth groups). They may threaten or depose corrupt local chiefs to gain access to the company rents and payment that flow from oil operations in their territories. Corporate practice and the struggle over oil rents to chiefs has contributed directly to the longstanding inheritance and dynastic conflicts that have so destabilized kingdoms in Nembe, Bonny, Opobo and Okirika, and more generally to what can only be described as a crisis of chiefly rule across the oil fields. The fact that so many chieftainships are ferociously contested and that in some cases chiefs and elders are routinely thrown from office, publicly abused or flogged by militant youth—actions unthinkable three decades ago—is an indication of how the old system is being replaced by the Niger Delta's own version of vigilante rule and mafia-style privatized violence.[41]

By almost any measure of social achievement, the core producing oil states are a calamity. The United Nations, in the most systematic account of development trends, estimates that between 1996 and 2002 the human development indices actually fell in the core oil-producing states. Literacy rates in the core states are barely 40 percent; the proportion of primary[42] school children enrolled is, according to a Niger Delta Environmental Survey (NDES), 39 percent. The decree of decrepitude in primary schools—whether in Port Harcourt or the further reaches of the riverine areas—is simply staggering: no desks, no teaching materials, no teachers and not unusually, the ratio of doctor to population is 1:27,000 in Delta State (and 1:282,000 in some of the local government areas in Southern Ijaw, Bayelsa State) There is one secondary health care facility for every 131,000 people serving an area of 583 square km. Electricity is a running joke. Outside of the urban areas only 20 percent of settlements are linked to a national grid that does not function in any meaningful sense. A striking discovery of the UN was the fact that local government areas with oil facilities are more likely to have significantly higher indices of human poverty.

Staggering under economic impacts, poor health and broken promises communities have little recourse under the Nigerian legal system. They are afraid to sue for cleanup and compensation because history shows that oil companies will appeal repeatedly until the plaintiffs run out of money, give up or die. Because access to justice is often denied victims of human rights abuses in Nigeria, many see other routes, including violence, as their only means to obtain some form of redress. There is a huge disparity of monetary resources available to oil companies for defense and local communities' resources to bring complaints. Legal costs are high, procedures lengthy and administration and lawyers' fees are substantial. This combination prevents communities and individuals from considering litigation and appeals processes that could financially ruin them. Whereas aggrieved individuals and communities have little, if any, legal aid to pursue their cases, oil companies can afford the best lawyers and can depend on the legal expertise of their parent companies. Widespread allegations of corruption in the judicial system also serve to deter attempts at legal action.

Having lost their traditional subsistent lifestyle to pollution and other drastic changes in their immediate environment, many oil producing communities are now forced to buy their food. This puts local people at an immediate disadvantage in comparison to the purchasing power of salaried company employees, many of whom come from other parts of Nigeria or from abroad. In the Niger Delta oil exploration has reportedly led to the loss of the fish population along the coast: fishing is available only to those who can afford large boat engines and trawlers to venture into the high seas. The rest of the population must buy "ice fish" (frozen fish) from commercial fishermen, a practice totally unknown a few years ago. Because market prices are constantly on the rise, many villagers go without fish.[43]

The diminished productivity and viability of local economies owing to the environmental and social degradation caused by oil exploitation has affected the lives of women in unique ways. The rights of women have been violated by oil companies in many ways. Women are no longer able to provide for their families by performing traditional roles of farming and trading in agricultural and other goods. Faced by such devastating economic circumstances, many women are forced to turn to prostitution as a means of survival. This situation is aptly captured by Joi Yowika who observed that "sex industry in the Niger Delta is directly linked to the oil industry, because it is oil company employees and the employees of oil-related service companies that patronize prostitutes. As a result prostitution is rampant in oil producing communities and in cities where oil workers reside."[44] This point is further buttressed by the commission of Nobel Laureates on Peace, Equity and Development in the Niger Delta Region thus:

> According to the United Kingdom's Department for international Development, ìNigeria has some of the worst social Indicators In the world.î For example, the World Bank estimates that 92.4% of Nigerian women are living on less than $2/day in 2006.
>
> Many people are migrating from their villages to Port Harcourt and other urban areas. The Niger Delta Region has the highest incidence of HIV/AIDS in Nigeria; prostitution and teenage pregnancies are on the rise. There is also spiral of hooliganism, robbery and spontaneous violence.[45]

This is similar to what happened to local communities along the Mackenzie Valley and Alaska pipelines. The same phenomenon has happened with all the "boom towns" everywhere throughout history: for example, the gold rush towns of California and the Klondike, or the silver towns of Nevada. These historical analogies of unrestricted greed and corruption are legion, and the consequences for women are always the same.

In addition, there is hardly any village in the area where "fatherless" children are not common. Frequently, abandoned children are common as women cannot trace the fathers of their children because as soon as the oil workers move to another location, they never return. Apart from inflicting a social cost on the women, the society in general has been traditionally "polluted" because "bastards" are taboo in the communities. Cases of rape, adultery and fornication have become so common that sacrifices offered to the ethical divinity are no longer cleansing the land of the immoral acts of the oil workers. Hence, the traditional culture has been ruined along with everything else.[46]

Within the labor force, the oil companies discriminated against locals. This leads to a situation where jobless youths migrate to towns and cities, and they abandon their families in the village in search of work. Ironically, in the towns and cities, there are no jobs for the migrating unskilled villagers because the oil money has not been invested there but in other major towns and cities outside the region or invested in the infrastructure from which the region hardly benefited.

A major source of concern and exasperation for people within and out of the Niger Delta is the widely perceived insensitivity of the political leaders of the area to the deteriorating living conditions of their own people. Oil wealth can be, and often is, ill managed. Oil-producing states are among the most corrupt and venal anywhere.[47] Corruption undermines economic growth, jeopardizes financial stability and undermines confidence in government institutions. It leads to criminality that has corrosive effects on governance and the rule of law. Federal and state governments have stolen or wasted $380 billion since the country became independent in 1960. The culture of corruption is pervasive, affecting state government as well as local officials: 31 of the 36 state governors are currently under investigation for crimes of corruption. The fact that Nigeria's coffers are awash in oil revenue further fuels corruption by the national and local political elite.[48]

The symbiotic relationship between multinational companies and communities in the Niger Delta behooves companies to provide community development projects in turn for a peaceful operating environment for the communities. But in many cases, companies do not deliver what they promised and have provoked tension and community protests. Organized protest and activism by affected communities are regularly met with excessive military repression, sometimes ending in the death and injury of unarmed civilians and the razing of whole communities. For instance, in November 1999, when 12 policemen were murdered by an armed gang in Odi Town in oil-rich Bayelsa State, President Obasanjo ordered in the security forces. The town was razed and hundreds of people were said to have been killed, apparently in reprisal for the murder.

Although some company-funded projects functioned, others have either been poorly run and accessible to a few communities only, or companies have awarded contracts and benefits in an arbitrary manner that perpetuates discrimination and marginalization. For instance, Odioma, Nembe-Bassambiri and Obioku communities of Bayelsa State have been competing for ownership and control of Obioku on the Santa Barbara River where Shell Nigeria has been prospecting for oil since 1988, in expectation of royalties, jobs and contracts. Although, Shell recognized Obioku and Bassambiri communities as the landowners, the company was compelled to withdraw from the area in January 2005 after a group of young men claimed Odioma's ownership of the land and demanded that operations stop.[49]

Whereas oil is a source of profit to the oil multinationals, the oil-rich environment is fundamental to the survival and reproduction of the oil minorities and the local host communities. Due to the fragile ecology of the Niger Delta, oil production has the impact of upsetting the delicate balance between land, water and life. There is no doubt that the clash of interests between the profit motives of the oil multinationals and the rights of oil minorities remain a sore point, and has further fuelled the feeling of the oil minorities that they are being marginalized and unjustly punished on account of their small size. Such feelings directly question their status in Nigeria's federalism and the liability of the state to protect their interests vis-à-vis the exploitative oil multinationals. It is this loss of faith that poses some danger to the Nigerian state as presently constituted.[50]

CONCLUSION

As the sixth largest producer in OPEC, Nigeria contributes substantially to the global oil market. The sale of crude petroleum in the last five decades has brought a phenomenal change in the country's economy. Paradoxically, the oil industry, which has brought development to many parts of Nigeria, has become the most environmentally deleterious enterprise to the people of oil-producing communities whose existence is now threatened by poverty, environmental pollution, a lack of infrastructure, unemployment and neglect by the Nigerian government.

Efforts to get the oil companies and government to address the seething anger of the people of the Niger Delta have subsequently met with failures, as such efforts are not far reaching and are merely palliative in nature. Thus, the efforts are just addressing the symptoms of the problems, not the disease that caused the problem. This severe economic neglect and social exclusion has resulted in deepening frustration and social unrest in the Niger Delta. By 1998, the Niger Delta region degenerated into a lawless zone and communities frequently engaged, with little provocation, in destructive intercommunity and intracommunity unrest, the wanton destruction of oil pipelines and the abduction of foreign nationals for ransom. Nigeria's major oil producers lost millions of barrels of oil due to forced stoppages by protesting villagers in oil-producing areas.

In view of the strategic importance of the Niger Delta to the country, profound changes will be required if there is to be lasting peace.

The Nigerian government should guarantee that the operations of oil industries are carried out in ways that can safeguard against environmental degradation and promote sustainable development, welfare, equity and justice. In addition the government should on its part respect citizens' rights and consider derivation principle for oil producing states, provide essential social infrastructure, create job opportunities for both educated and skilled members of the community as well as alternative employment for farmers and fishermen whose traditional means of livelihood have been eroded by oil production as well as micro-credit schemes for women.

NOTES

1. Dele Agekamah, "A Region and its Hidden Treasures," *The Capitol* 1, no. 2, (2006): 34.
2. "The Elie Wiesel Foundation for Humanity, Commission of Noble Laureates on Peace, Equity and Development in the Niger Delta Region of Nigeria," in *Curse of the Black Gold. Fifty Years of Oil in the Niger Delta,* ed. Michael Watts (Brooklyn, N.Y.: Power House Books, 2006), 218.
3. Agekamah, "A Region and its Hidden Treasures," 34–35.
4. Ike Okonta and Oronto Douglas, *Where Vultures Feast: Shell, Human Rights, and Oil* (London, New York: Verso, 2003), 19.

5. Okonta and Douglas, *Where Vultures Feast,* 19.
6. Ibid., 64.
7. Oroto Douglas, Dimieari Von Kemedi, Ike Okonta, and Michael J. Watts, "Alienation and Militancy in the Niger Delta: Petroleum, Politics, and Democracy in Nigeria," in *The Quest for Environmental Justice: Human Rights and the Politics of Pollution,* ed. Robert D. Bullard (San Francisco, Calif.: Sierra Club Books, 2005), 241–242.
8. Segun Gbadegesin, "Multinational Corporations, Developed Nations, and Environmental Racism: Toxic Waste, Exploration, and Eco-Catastrophe," in *Faces of Environmental Racism: Confronting Issues of Global Justice,* 2nd ed., ed. Laura Westra and Bill E. Lawson (Oxford: Rowman and Littlefield Publishers, Inc., 2001), 195.
9. Toyin Falola and Ann Genova, *The Politics of the Global Oil Industry: An Introduction* (Westport, Conn.: Praeger, 2005), 104–105.
10. Martin O'Malley, *The Past and Future Land: An Account of the Berger Inquiry into the Mackenzie Valley Pipeline* (Toronto: P. Martin Associates, 1976), 86.
11. "Oil Industry has brought poverty and Pollution to Niger Delta," http://www.amnesty.org/. Accessed April 5, 2010.
12. Michael Fleshman, "Report on the Crisis in Nigeria," http://www.prairient.org/acas/alerts/nigeria/nigeria1.html. Accessed April 5, 2010.
13. Bronwen Manby, *The Price of Oil* (New York: Human Rights Watch, 1989), http://www.hrw.org/reports/1999/nigeria. Accessed May 4, 2010.
14. Dokubo Asari, "Corroding the Soul of a Nation," in Nicholas Shaxson, *Poisoned Wells: The Dirty Politics of African Oil* (New York: Palgrave Macmillan, 2007), 201.
15. Leo Akpatu, "Resource-Based Conflicts: Challenges of Oil Extraction in Nigeria" (paper presented at a European Conference Hosted by the German EU Council Presidency, Berlin, March 29–30, 2007), 13.
16. Henry Clark et al., "Oil for Nothing: Multinational Corporations, Environmental Destruction, Death and Impunity in the Niger Delta," *A US Non-governmental Delegation Trip Report* (September 6–20, 1999), 9.
17. Beverly Wright, "Living and Dying in Louisiana's 'Cancer Alley,'" in *The Quest for Environmental Justice,* ed. Robert Bullard, 103.
18. Mathew Yeomans, *Oil: Anatomy of an Industry* (New York: The New York Press, 2004), 81–82.
19. Tom O'Neill, "Curse of the Black Gold: Hope and Betrayal in the Niger Delta," *National Geographic* 211, no. 2 (February 2007), 97.
20. Ibid., 97.
21. E. J. C. Duru, *Oil Multinationals and the Niger Delta Crisis: Issues and Perspectives* (Owerri, Nigeria: African Scholars Publishing Company, 1999), 92–93.
22. Gbadegesin, "Multinational Corporations, Developed Nations, and Environmental Racism," 197.
23. Falola and Genova, *The Politics of the Global Oil Industry,* 197.
24. Yinka Omorogbe, *Oil and Gas Law in Nigeria* (Lagos, Nigeria: Malthouse Press Ltd., 2003), 58.
25. "Gas Flaring in Nigeria: A Human Rights, Environmental and Economic Monstrosity" Available at http: www.foe.co.uk/resource/reports/gas_flaring_nigeria. Accessed April 11, 2010.
26. Tayo Akeem Yusuf, "Oil and Gas Industry Activity and the Niger Delta Environment: The Role of Environmental Law," http://www.nigeriansinamerica.com/articles/3191/1/Oil-amp-Gas-Industry-Activity. Accessed February 5, 2009.

27. Clark et al., "Oil for Nothing," 5.
28. Duru, *Oil Multinational and the Niger Delta Crisis,* 94.
29. Watts, "The Elie Wiesel Foundation for Humanity," 218.
30. Gbadegesin, "Multinational Corporations," 196.
31. Yusuf, "Oil and Gas Industry Activity and the Niger Delta Environment," 4.
32. See Yinka Omorogbe, "An Appraisal of Nigerian Natural Gas Legislation," *Oil and Gas Law and Taxation Review* 4, no. 2 (1985), 51.
33. Johan Fabricius, *East Indies Episode (Shell war Achievement 3)* (London: Shell Petroleum Company, 1951).
34. Yusuf, "Oil and Gas Industry Activity and the Niger Delta Environment," 4.
35. Watts, "The Elie Wiesel Foundation for Humanity," 218.
36. Yeomans, *Oil: Anatomy of an Industry*, 81–82.
37. Watts, "Sweet and Sour," 44.
38. United Nations Development Program, "Niger Delta Human Development Report" (Abuja, Nigeria, 2006). See also Watts and Kashi, *Curse of the Black Gold* (Brooklyn, N.Y.: Powerhouse Books, 2008), 1.
39. Tom O'Neill, "Curse of the Black Gold," 97.
40. Asari Dokubo, "Corroding the Soul of a Nation," in *Poisoned Wells: The Dirty Politics of African Oil,* ed. Nicholas Shaxon (New York: Palgrave Macmillan, 2007), 197.
41. Watts, "Sweet and Sour," 46.
42. Ibid., 44.
43. Clark et al., "Oil for Nothing," 9.
44. Ibid., 10.
45. Watts, "The Elie Wiesel Foundation for Humanity," 218.
46. Duru, *Oil Multinational and the Niger Delta Crisis,* 95.
47. Watts, "Sweet and Sour," 40.
48. Ibid., 218.
49. Consult Amnesty International, "Nigeria Ten years on: Justice and Violence haunt the Oil Delta." http://web.amnesty.org/library/Index/ENGAFR 440222005. Accessed February 5, 2009.
50. Cyril I. Obi, "The Impact of Oil on Nigeria's Revenue Allocation System: Problems and Prospects for National Reconstruction," in *Federalism and Political Restructuring in Nigeria,* ed. Kunle Amuwo et al. (Ibadan, Nigeria: Spectrum Books Limited, 1998), 271.

6 Illiberal State, Oil Expropriation and Environmental Degradation in Nigeria's Niger Delta Region

Dele Ashiru

INTRODUCTION

Ominous signs that the Niger Delta environment was in peril became manifest after the discovery of oil in 1956 at Oloibiri, a community in present day Bayelsa State. The agitation of the minorities and especially the people of this area against the usurpation of their land for oil exploration and fears of discrimination and marginalization in the political and economic spheres of life after independence led to the setting up of the Sir Henry Willink Commission preparatory to Nigeria's independence. At the close of deliberations, the Willink Commission acknowledged the genuineness of the minorities' fears and found evidence of possible marginalization after independence, but rather than take fundamental structural and institutional steps to allay these fears, such as the creation of states to accommodate ethnic minorities, the Nigerian illiberal State in connivance with the Commission, acting out the script of the colonial government, made feeble recommendations such as entrenching in the emerging Independence Constitution a bill of rights to protect the people of the Niger Delta region and other ethnic minorities in the country with similar fears and concerns. Although the Willink Commission noted the present and future danger the discovery of oil portended for the Niger Delta people and their environment, it merely recommended the creation of a special commission to address the peculiar environmental problems envisioned from oil exploration in the region.[1]

When independence was eventually achieved in 1960, the ruling elites, as will be shown later, found it convenient to inherit the illiberal state from colonial rule without an attempt to transform it. There is little wonder, therefore, that shortly after independence and several decades since, all the Nigerian illiberal state can show for the stupendous wealth received from oil revenue accrued from the Niger Delta area are increasing unemployment, an uncontrollable spread of preventable diseases, unbelievable squalor, decaying infrastructures, an epileptic power supply, inconclusive and often violent electoral contestations producing illegitimate rulers who lacked respect for human life and civil liberties, wide inequalities and the criminal neglect of the environment, which resulted in environmental degradation of the Niger Delta area from where the bulk of the oil revenue comes.

This chapter intends to demonstrate that the illiberal nature of state formation in Nigeria compels it to act not in the interest of the people following the logic of social contract theories but in the interest of the colonial and neo-colonial powers represented by ruling elites and multinational oil corporations who forcefully glued the people together purely for economic exploitation and pillage in the first instance. The central argument is that the illiberal nature of state formation, the fact that Nigeria was not a result of any dialogue, consensus building by the people nor was it a product of any democratic engineering deriving from the consent of the people results not only in the undemocratic nature of the polity but in the criminal oil expropriation and the consequent environmental degradation of the Niger Delta region. The rest of the chapter is divided into four parts including conceptual and theoretical concerns, the Niger Delta region, oil expropriation and agitations for self-determination and in conclusion what is to be done.

THEORETICAL AND CONCEPTUAL CONCERNS

> and this that which begins and actually constitutes any political society is nothing but the consent of any number of freemen capable of a majority to unite and incorporate into such a society. And that is that and that only which did or could give beginning to any lawful government in the world.[2]

In its pristine form, liberalism is associated with the belief in the rights of individuals to life, liberty and property and also the right of a people to national self-determination and democratic government. Liberalism as a political theory is often associated with John Locke, which extols the virtues of constitutionalism, stability, freedom, consent and tolerance. In Locke's exposition of liberty from which liberalism derived, liberty at the individual level "was not freedom to do what one chooses in its literal sense, but the duty to act within the limits of the laws of nature, while freedom presupposes order which could only be made possible within the framework of law."[3] At the level of the state, it must derive from the consent of the people, which they freely gave to those in authority. These individuals in authority must therefore act within the ambiance of the law, for to do otherwise will make it illiberal. Espousing the characteristics of a good state, Locke argues that it must exist for the people who formed it, and not vice versa. The state had to be based on the consent of the people subject to the constitution and the rule of law. It would be limited because its powers were derived from the people and were held in trust.[4]

Advancing the liberal argument further, Emmanuel Kant's exposition on a threefold categorization of rights becomes very useful. According to Kant, liberalism advocates freedom from arbitrary authority, which includes freedom of conscience, free press and free speech, equality before the law and the right to hold and exchange property without fear of arbitrary seizure of it. These are regarded as "negative freedoms." "Positive freedoms," on the other hand, extol social and economic rights, such as equality of opportunity in education and rights to health care as well as gainful employment. All of these are necessary for effective self-expression and participation in the affairs of the state. The third categorization has to do with the right to democratic participation without which the two other categories of freedoms and rights cannot be effectively guaranteed.[5] The critical questions to ask are: what is a state? Did the Nigerian state at inception derive from the consent of the people? Does the state engender democratic participation? Can the government instituted by such a state be constitutional and operate within the ambiance of the law? Do the people at any point consent to live together as one entity? Could a state formed in the manner of Nigeria be a liberal one? Can such a state act in the interest of her people?

No concept seems to be so central to political discourse than the state but there is yet to be a universally acceptable definition of a state. Whereas some scholars see the state as essentially a class structure, others view it as an organization that goes beyond class differentiation but attends to general needs across societal classes. What is incontrovertible is that the state remains the most important and distinctive institution in any society. Yet, the state is not a material object; it is a mere conceptual abstraction.[6] It is that organization that gives color and direction to society, although Omeje once cautioned that "a clear cut demarcation between the state and society can be analytically tricky."[7] In spite of this, there are basically three distinctive ways through which the state can be understood.

The three distinctive ways in which a state can be defined include the institutional level, the functional level and the relational level. Ismail captures these three conceptions of the state when he submits that

> on the institutional side, the state is a set of interrelated governmental institutions responsible for making rules, controlling and regulating social relations; functionally, the state is a set of public institutions established to accomplish specific objectives such as the provision of security and welfare for people under its Juridical domain; and relationally, the state is a set of relationship[s] and interactions among social classes and groups that is maintained, organized and regulated by political power.[8]

Of the three conceptions of the state, the relational aspect as enumerated above is of crucial importance to understanding the nature and character of any state because it is based on the relationship and interaction among social classes and groups, which is maintained, organized and regulated by political power. When political power is unorganized and unregulated it negates the very essence of state especially when the state is understood from the perspective of the social contract theorists, where the people freely consent to surrender their rights to a sovereign who exercises those rights on behalf of all in order to safeguard them.

The question therefore is for what purpose is the state established? Three primary purposes for which the state is created are the duty of protecting society from violence and invasion of other independent societies, the duty of protecting subjects from injustice and the duty of erecting and maintaining certain public works and public institutions.[9] The extent to which a state can do this depends on the history, nature and character of the state and the ruling elites it produces. What then are the history, nature and character of the Nigerian state and its ruling elites?

The reality that the Nigerian state is a colonial creation is incontrovertible. Suffice to say, that before the amalgamation of 1914, the geographical entity now known as Nigeria was populated by diverse ethnic groups such as the Oyo Empire, the Benin Kingdom, the Sokoto caliphate and scores of minority ethnic groups in the middle belt and the Niger Delta including the Igbo communities. Whereas it is not sufficient to assert that Nigeria was a colonial creation, what is of analytical importance to this study is the nature of the colonial state that continued to have far-reaching effects on the post-colonial state and the nature of the politics within the ruling elites in Nigeria since independence.[10]

The colonial state was a law and order state with absolute and arbitrary powers at the political and economic levels.[11] At the political level, the arbitrary use of power deprived it of legitimacy, as there were constant antagonisms between the state and the people. This can be said to have been the basis of the crises of legitimate governance, which have trailed the state from independence.[12] At the economic level, the colonial state was essentially an extraction enterprise in conformity with the logic of colonial exploitation and pillage, which also resulted in the forceful incorporation of the country's economy into the world's capitalist system as a weak and junior partner.

It bears no emphasis that this had debilitating effects on the country's political economy because the forceful incorporation disarticulated the economy through the introduction of cash crop farming, which relied heavily on capital and technology from the metropole. This ensured the dominance of foreign capital and technology in the economy because indigenous methods were no longer required for agricultural production. Once independence was achieved, the state and its oppressive apparatuses were inherited without any desire to change or transform it. All these factors were coupled with ruling elites at independence that lacked any productive base as well as entrepreneurial abilities owing to an overreliance on and dominance of foreign enterprises in the economy and therefore, reliance on the state as a means of production for elites' unbridled material acquisition and primitive accumulation. Their activities thus became parasitic and predatory to the state.

The consequence of the foregoing was the high premium placed on the capture and retention of political power to have unfettered access to the state and the privileges it confers. Given this situation, the struggle for political power was conducted in a war-like manner. Nigerian politics from then on knows neither rationality nor compromise, but expediency. Ake sums this up as thus:

> In the absence of autonomizing mechanisms in the post-colonial state, the resources of physical coercion become the tools of particular groups, especially the hegemonic factions of the ruling class. . . . So, we have essentially relations of raw power in which right tends to be coextensive with power and security depends on the control of power. The struggle for power then is everything and is pursued by every means.[13]

This regrettably remained the bane of Nigeria's politics since independence, as virtually all the elections conducted in the post-colonial history of Nigeria have been characterized by electoral malpractices and violence of an unprecedented proportion. The ruling elites brought this about by flawed electoral processes and were usually incapable of running the affairs of state effectively; apart from this the overreliance on oil rents in terms of the mono-cultural nature of the economy and the monopoly of oil technology by multinational corporations put the state under foreign influence, especially Western capital.[14] The state is therefore weak and compromised in its dealings with multinational oil corporations as it affects their explorative activities in the Niger Delta region.

For our purpose here, oil expropriation is used to denote the forceful usurpation, violation of land and the entire environment of the Niger Delta for oil exploration and exploitation purposes by the Nigerian state and its international oil multinational collaborators ably supported by their respective home governments. This was done through the enactment of laws that vested in the Federal Government the ownership of all mineral resources located anywhere in the federation. The Minerals Act of 1958 was enacted to amend and consolidate all the existing legislations relating to mines and minerals in the country. It also vested ownership of mineral resources including oil and petroleum in the Federal Government. Indeed, the 1959 Petroleum Profit Tax Ordinance provided that oil profit would be shared between the state and oil companies equally without any due consideration for the people and the environment where oil is explored. It is also disheartening to note that both the Independence and Republican constitutions of 1960 and 1963, respectively, gave the federal government exclusive rights and powers to legislate on "mines and minerals including oilfields, oil mining, geographical surveys and natural gas."[15] Under the military, the Petroleum Decree No. 51 of 1969 further consolidated the federal government's hold on legislations that concerned oil and petroleum resources in the country.

All the above legislations, it must be noted, were enacted in complete disregard for the people whose natural habitat and topography are violently violated by the capital intensive nature of oil production. This could, however, not have been otherwise given the nature of the state and the character of her ruling elites depicted above. Also, as noted by Obi,[16] to gain access to the local communities for oil exploration, the oil multinational corporations entered into a pact with the state either by the state granting the multinationals oil concessions as was the case between 1914 and 1969 or through a Joint Venture Agreement, Production Sharing Contract or Risk Sharing Contract.

Within this broad framework, the Nigerian state issues various categories of licenses such as the Oil Exploration License (OEL), the Oil Prospecting License (OPL) and the Oil Mining Lease (OML), which gives multinational corporations unfettered access to the Niger Delta for unbridled oil exploration with devastating consequences on the area's environment. However, of specific importance to our analysis in this chapter is the OML, which apart from entitling the holder to prospect for and dispose of whatever petroleum mined in the area allocated for a period of twenty years subject to renewal; the holder of an OML license is obliged to respect the rights of any person or persons in lawful occupation of land in the area of operation by entering into negotiation for payment of fair and adequate compensation for the disturbance of their surface rights.

In addition, the owner of the lease must ensure that he does not deprive any land, village, houses or watering place for cattle of a reasonable supply of water or interfere with any rights enjoyed by any person. Accordingly, "the lessee must be ready to pay adequate compensation for things such as economic trees, structures affixed to land object etc., besides, he is debarred from tampering with sacred lands, burial grounds, public installations, markets etc within a specific distance from his area of operation. In sum, the lessee must be mindful of the requirements of conservation and sustainable development, all of which have become concretized in good oil field practices."[17] In the same vein, Oyebode further notes that the lessee must exercise due diligence by taking all practical means "to prevent the escape of petroleum into any water, well, spring, stream, river, lake, reservoir, estuary or harbor; and to cause as little damage as possible to the surface of the relevant area and to the trees, crops, buildings, structures and other property there on."[18]

Do oil companies observe all of these principles in their operations in the Niger Delta? The answer is of course a categorical no. Indeed, it is the failure of the relevant agencies of the state, such as the Nigeria National Petroleum Corporation (NNPC), Department of Petroleum Resources (DPR), Federal Ministry of Environment (FME) etc., on account of their weakness and acting in connivance with oil multinational corporations and their inability to enforce the laws enacted for environmental protection, that have resulted in the environmental degradation of the Niger Delta region as manifested in air and water pollution, through unbridled gas flaring, oil spillages, blow-outs and discharge of industrial waste.

Beyond the issuance of operating licenses and the oil mining lease is the issue of the Land Use Decree Promulgated in 1978. This Act has remained as controversial and contentious today in the Niger Delta region as it has been since it was promulgated by General Olusegun Obasanjo's military administration. The Land Use Decree vested in the state the ownership of lands and even provides that occupancy can be revoked if the land is required for mining or oil-processing activities. Through this decree, which later found its way into the 1999 Constitution under section 315(D) as the Land Use Act,[19] the people of the Niger Delta have been expropriated from their land and are excluded from ownership of their land and the resources that accrue from it. As Obi observed, "exclusion breeds resentment of the powerful extractor and raised the issue of the immorality of expropriation . . . where the people of the Niger Delta have suffered government neglect over the years."[20]

Nothing lends more credence to the illiberal nature of the Nigerian state than its reaction to the agitations of the people of the Niger Delta region to have a fair share of the rents accruable to the state from the sale of oil and gas, as well as draw the attention of the state and oil multinationals to the severe damage being done to the environment through oil exploration. These political, socioeconomic and environmental neglects have intensified the call from the people of the Niger Delta for resource control, political inclusion and self-determination.

It is needless to say that the reaction of the state has been unbridled oppression, exploitation, repression and outright militarization of the region, resulting in rape, killing, arson and state-sponsored violence. Evidence of state repression can be found in the Umuechem massacre of 1990, the military campaigns of the Rivers State Internal Security Task Force against the Ogoni (1993–1998), the case of Parable village shooting incident in May 1998 and the invasion of the Niger Delta in December 1988 and January 1999 by security forces following the Kaiama Declaration by the Ijaw Youth Council.[21] The manifestations of the illiberal nature of the Nigerian State in its relationship with the Niger Delta are that the oil producing communities have borne the brunt of oil production and gained very little if anything for their pains and sacrifices; apart from this, the politics of power and revenue sharing has been consistently designed and implemented to the disadvantage of the oil-producing areas, thereby depriving them opportunity for growth, development and progress. The state and its apparatuses, controlled by elements from non-minority and non-oil-producing areas, are responsible for the overcentralization of power in the federal government. The central government has relied on violence and all sorts of human rights abuses to intimidate and control the inhabitants of oil producing communities and their respective organizations. Also, the neo-colonial state has forged a powerful alliance with oil companies in the degradation of the environment and the suppression of the voices of people and organizations in oil producing areas.[22]

ENVIRONMENTAL DEGRADATION AND THE STRUGGLE FOR SELF-DETERMINATION

> (This) region which contributes enormously to the nation's exchange earnings is suffering from administrative neglect, crumbling social infrastructure and services, high unemployment, social deprivation, abject poverty, filth, squalor, and endemic conflict.[23]

Nothing sums up the situation in the Niger Delta region better than the caption quoted above. The region described as one of the largest wetlands in the world is estimated to cover an area of about 70,000 square kilometers, which consists of a number of ecological zones that range from sandy coastal ridge, permanent and seasonal swamps, saline mangrove as well as forests. Indeed, the Niger Delta Environmental Survey (NDES) of 1997 defined the Niger Delta region in four cartographic ways, which are: the natural Niger Delta, the political definition of the Niger Delta, the Oil Mineral Producing Areas Development Commission's (OMPADEC), definition and the operational definition for the purpose of NDES. For our own purpose, the natural Niger Delta that emphasizes the geographical location will suffice. Thus, the Niger Delta is described as "the low lying actuate deltaic plain with a northern apex a little South of Onitsha at Aboh (5033'44"N6031'38"E), a western apex by the estuary of the Benin River (5044'11"N5003'49"E) and an eastern apex at Imo River estuary (4027'16"N, 7035'27E). The Southern most part is believed to be at palm point (4016'22"N6005'27"E), South of Akassa at the estuary of Nun River of the two major distributaries of Niger River. It is estimated to cover almost 26,000 km^2, approximately 2.8% of the land area of Nigeria. The axis of the Delta is in a NNE-SWS direction."[24]

The Niger Delta territory is inhabited by the Ijaws, Ogoni, Itsekiris, Urhobos, Ikweres, Andoni, Ndokwa, Kwale, Ilajes and Isokos with a population of over ten million people living in separate communities with fishing and subsistence agriculture as the major occupations of the people before the discovery of oil. Today, these communities are found in Bayelsa, Rivers, Akwa Ibom, Ondo, Delta, Edo and Imo, albeit unwillingly, but through the oppressive, inelegant, criminal, unilateral and incessant history of arbitrary creation of states (subunits) in Nigeria.

As indicated above, the discovery of oil in Oloibiri now in Bayelsa state in 1956 led to the invasion of the Niger Delta area by scores of multinational oil corporations led by Shell Petroleum Development Company (SPDC). Through oil exploration activities and several legislations enacted for that purpose, the people were expropriated off their land while leaving their environment thoroughly and mindlessly degraded. Their land, waters, streams and creeks were despoiled through the reckless activities of these multinationals in the process of oil extraction. Their sources of livelihood were lost owing to erosion by ships carrying heavy duty equipment to oil locations and by the canalization of salt water into fresh water to enable navigation for oil activities.

Consequently, there are many environmental problems confronting the Niger Delta. In addition to the difficult terrain, which is a large problem, there are other problems arising predominantly from the activities involved in oil prospecting and production. Such problems include air, water, land and chemical pollutions as the main sources of environmental degradation in the Niger Delta.[25] In terms of air pollution, oil companies indiscriminately flare gas in the Niger Delta that emits toxic chemicals into the atmosphere. It is estimated that about 76% of all natural gas produced in Nigeria is flared and this is the highest in the world. This process produces about 35 million tons of carbon dioxide and about 12 million tons of methane annually with serious effects on the depletion of the ozone layer, thereby contributing to the greenhouse effect as well as acidic rain in the Niger Delta.[26] Nigeria flares about 24 billion cubic meters of gas annually, which accounts for 12.5% of the entire flared gas in the world, compared to 0.6% in the United States of America and 4.3% in the United Kingdom. In addition, Nigeria losses about 6 billion naira, daily, to gas flaring. It is also on record that oil companies have continued to force the Nigerian state to rescind on deadlines to end gas flares as initially agreed.[27]

Oil spillage is another form of environmental degradation in the Niger Delta. Oil production is said to account for more than 600 major spills every year.[28] In the event of spillage there are usually accusations and counteraccusations about sabotage between the communities and the oil-producing companies. For example, whereas local inhabitants demand compensation for oil spills, the oil companies in year 2000 claimed that 57 percent of oil spills were due to sabotage. The oil companies also insist that 63 percent of spills in 2003 were also due to sabotage,[29] whereas several millions of naira has also been paid by the oil companies as compensation to some host communities. The concern however, is that oil spillages changed the character of the land by polluting it to the detriment of living organisms. Such spillages are thus responsible for the destruction of farms, crops and vegetation, including wildlife; it has also impaired human health. Food items like plantain, cocoyam and yams change their natural colors on account of spills. Indeed, Inhabitants in the area have even complained that their food items smell and taste like kerosene when eaten.[30]

Apart from these environmental problems, the people themselves have nothing to show for the wealth derived from their territory other than political marginalization, exclusion, social deprivation, extreme poverty, unemployment and lack of infrastructural facilities such as schools, electricity, roads, drinkable water and hospitals. The foregoing analysis is aptly captured in the following words enshrined in the Kaima Declaration of 1988 by the Ijaw Youth Council:

> the unabating damage done to our fragile natural environment and to the health of our people is due to in the main, to uncontrolled exploration and exploitation of crude oil and natural gas which has led to spillages, uncontrolled gas flaring, the opening up of our forests to loggers, indiscriminate canalization, flooding, land subsistence, coastal erosion, earth tremors etc. Oil and gas are inexhaustible resources and the complete lack of concern for ecological rehabilitation in the light of the Oloibiri experience, is a sign of impending doom for the people of Ijaw land.[31]

Under the scenario related above the people of the Niger Delta, especially the youths, began to engage the state to get a fair share of the resources derivable from their soil. For example, more than 600 billion U.S. dollars was said to have accrued as rents from oil rents since oil was discovered in commercial quantity in Oloibiri, present day Bayelsa state. But, rather than increasing the revenue sharing formula that emphasized derivation at inception, the formula decreased from 100% in 1953 to 50% in 1960. It was further reduced to 45% in 1970 and 20% in 1978, 2% in 1982 and 1.5% in 1984 before it was raised to 13%, which is not being faithfully utilized in the development of the region. This sharp decline in the revenue and the continued environmental degradation prepared the stage for the current violent crisis in the Niger Delta.

Within this context, it can be argued that the injustices meted to the people of the Niger Delta and the non-equitable basis of Nigeria's federalism provided the logic to the conflicts that have engulfed the Niger Delta. However, the first known attempt to forcefully challenge the state was the secessionist activities of Isaac Adaka Boro, who declared the sovereignty of the Niger Delta People's Republic on the 23rd of February, 1966. He was later arrested with his other compatriots; they were tried, found guilty of treason and were summarily executed. It is instructive to note that three major phases are discernable in the people of the Niger Delta's engagement of the Nigerian state about these glaring injustices meted to this region.

The first phase, between the mid-1970s and early 1980s, was the period when the country received large amounts of revenue from oil and it is usually described as the "oil boom" period. During this period, as the country became more prosperous, the people of the region suffered neglect. The agitation of the people of Niger Delta at this period was thus limited to demands for compensation, damages, employment, infrastructural development and a better life for their people. During the second phase between mid-1980s and the mid-1990s, their agitation for compensation for acquired oil fields and the provision of basic amenities became more intense through the activities of youths such as protests, occupation of oil installation and disruption of exploration activities. At this stage, the operations of the oil companies became threatened and they had no choice but to call on the illiberal state for protection. Rather than discuss with the youths and people of the region ways toward finding solutions to the environmental problems and other related issues raised by the people, the state increased the militarization of the entire region and brutalized the people at the slightest provocation through its oppressive apparatuses and agencies.[32]

With the increase in militarization and brutalization, the youths in the Niger Delta also became more organized and radicalized to effectively engage the state. Thus, between the years 1990 and 1995, the struggle in the Niger Delta became very intense and internationalized through the activities of the Movement for the Survival of Ogoni People (MOSOP) led by Ken Saro-Wiwa. Ken Saro-Wiwa was arrested by the state in May 1994 on trumped up charges of felony and was killed. The killing of Saro-Wiwa seemed to be the last straw in the struggle for sociopolitical and economic emancipation of the people of the Niger Delta and marked the end of peaceful agitations by the people of the Niger Delta.

The third phase in the struggle for self-determination in the Niger Delta region is characterized by the acquisition of sophisticated arms and ammunition to violently confront the state and the multinational oil corporations. This confrontation has continually paralyzed oil exploration activities and shattered the peace and harmony for which the region was hitherto known for. This agitation and struggle of the Niger Delta people have led to protracted conflicts and violence in the Niger Delta resulting in hostage taking; kidnapping of expatriates, politicians and prominent citizens for ransom; political-assassination; and outright banditry. It is to be noted that the three phases are not mutually exclusive because these responses represent the reactions of the people to long years of injustice and marginalization of the people of the region. However, the distinctive feature of the third phase manifests in the quantum of instruments of violence at the disposal of some frustrated and disenchanted youths usually referred to as "militants" against the state. Indeed, "the conflict in the Niger Delta is the outburst of deep rooted frustration, helplessness and hopelessness by determined inhabitants fighting to right the wrongs embedded in the distorted Nigerian federalism."[33]

Apart from the use of military force by the illiberal state, there have also been some interventionist efforts through the setting up of one commission or the other. The earliest attempt was the setting up of the Niger Delta Development Board (NDDB) in 1962, following the recommendations of the Willink Commission. After this came the Niger-Delta River Basin Development Authority (NDRBDA), the Oil Minerals Producing Areas Development Commission (OMPADEC) established in 1992, the Niger Delta Development Commission (NDDC) established in 2000 and now the Ministry of the Niger Delta.

All of these interventionist agencies, except the newly established Ministry of the Niger Delta, have failed as a consequence of corruption through the criminal and reckless looting of public funds, inept and inefficient bureaucracy, crude and primitive accumulation of oil rents by the local elites as well as the culture of impunity usually exhibited by occupiers of public office acting in connivance with multinational oil corporations in total disregard of the people.

Specifically, these interventionist agencies failed because there is a general underestimation of the difficult terrain of the region by the leadership of the agencies. For example, the riverine areas are marshy and swampy. According to Governor Peter Odili of Rivers state, one of the oil-producing states, "what we (governments in the region) spend to do one kilometer of road could build 25 kilometers elsewhere in the country."[34] Also, undue political interference in the activities of the agencies through indiscriminate awards of contracts, which usually result in duplication of projects, creates inertia and detaches the people from the ownership of such projects. Apart from this, poor funding is also a reason why the agencies have failed. For example, among the 11 River Basin Development Authorities created in 1979, the Niger Delta River Basin Authority was the second least funded in the country. Also, between 1992 and 1996, only 13.2 billion naira was made available to OMPADEC to develop infrastructure in the region against a budgeted estimate of 85.5 billion naira. For the NDDC, only 43 billion naira was made available to it against the budgeted estimate of 75 billion naira at a time when the government was spending 50 billion naira to build a stadium in Abuja.[35] Despite the shortfall in funding, it is disheartening to note that the 'paltry' funds available to these agencies are also either misappropriated or outrightly embezzled through unbridled corruption that has become the rule rather than the exception in Nigeria. It is on record that the first two sole administrators of OMPADEC, Albert K. Horsfall and Professor Eric Opia, were dismissed on account of wanton corruption. Opia for instance was dismissed for his inability to account for 6.7 billion naira, about 80 million U.S. dollars,[36] and similar allegations trail most individuals who have served in the leadership of these agencies. Lack of an integrated development master plan, wastefulness and poor execution of projects without proper monitoring are additional major reasons why the agencies have failed.

WHAT IS TO BE DONE?

In line with the theoretical underpinnings of liberalism espoused in the second section of this chapter, John Locke, one of the social contract theorists, "categorically asserted that the government (acting on behalf of the state) could be altered, amended, changed or dissolved legitimately" in the event of any of the following situations:

> Whenever such a prince or single person established his own arbitrary will in place of the law; when the prince hindered the legislature from assembling in its due time or from acting freely, pursuant to those ends for which it was constituted; when by the arbitrary power of the prince, the elections and the ways of elections were altered without the consent, and contrary to the common interest of the people; the delivery of the people into the subjection of foreign power, either by the Prince or by the legislature; the person who had the supreme executive power neglect[s] laws already enacted and could not be executed.[37]

From the foregoing analysis, the Nigerian state has violated all of the conditions listed above. Thus, in line with this logic, coupled with the abundant evidence that the Nigerian state has consistently subverted the very essence and *raison d'etre* of a state, the state needs to be renegotiated. This has become more compelling than ever because the Nigerian state in its present form is weak, dependent, predatory, parasitic, inefficient and irresponsible. It is in this connection that we join the call for the convocation of a Sovereign National Conference (SNC) of all ethic nationalities within the Nigerian federation to deliberate on all national questions for the purpose of shedding the state of its colonial vestiges and its illiberal characteristics. This is the irreducible minimum condition to redress not only the injustices in the Niger Delta but the overall developmental challenges afflicting Nigeria.

Further, as part of negotiations at the proposed conference, resource control, political inclusion by, as well as the principle of self-determination must form the bedrock of negotiations. It is believed that it goes without saying that the decisions reached at this "Sovereign" dialogue should not be subjected to any tinkering either during the conference or after it, as has been the practice in the past by the few ruling elites and predators who continued to hold the state at its jugular. When this is done, it would then pave the way for the success of all other measures that would "halt the environmental degradation which oil exploration has foisted on the (region) . . . and return the Niger Delta to sound ecological health so that it continues to provide sustenance to present and coming generations."[38]

It is also expected that the outcome of the Sovereign Conference would guarantee the autonomy of the state through popular democratic empowerment, in which case the peoples' vote would count. The current charade called democratization, driven by the logic of liberal democracy where people vote without choosing, must give way to a truly participatory electoral process where the people can freely elect those who would run the affairs of state and take decisions on their behalf. The 'first- past- the- post' electoral system inherited from the colonial masters is not suitable for a sharply divided society like Nigeria. The system is not only nonconsensual but ensures that the winner wins everything and the loser losses everything; little wonder, therefore, that the history of elections in Nigeria has been characterized by electoral malpractices such as rigging, underage and multiple voting, stuffing of the ballot boxes, false declaration of election results and outright violence, all of which must give way before the state can become autonomous. In this regard, the report of the Justice Mohammadu Uwais panel on electoral reforms set up by Umaru Yar'adua administration, which removed all the clogs in the wheel of progress of the Electoral Commission especially in terms of autonomy and independence, should be implemented without further delay. Only then can the state be free from the shackles of a few privileged individuals and their foreign collaborators.

Once the state becomes autonomous from the few privileged ruling elite and their multinational collaborators, its relevant agencies such as the Department of Petroleum Resources (DPR), National Oil Spill Detection and Remediation Agency (NOSDRA), the Niger Delta Development Commission (NDDC) and the recently created Ministry of the Niger Delta, will become emboldened to enforce both national and international laws guiding the operations of multinational corporations in oil exploration as well as safeguarding the environment from further violation, abuse and degradation.

Consequently, adaptation, which refers to finding ways of coping with the problems climatic change has caused, and mitigation, which is concerned with how to stop further warming up of the globe through oil spillages and gas flaring, must be the bedrock on which government policies and actions must be based in the Niger Delta. Oil multinational corporations involved in oil exploration activities that affect climatic change should be compelled to adopt global best practices in oil exploration and environmental preservation in line with the dictates of the Kiyoto Protocol and the United Nations Framework on Climatic Change Convention. Although the United States of America (USA) is yet to sign the Kiyoto protocol, it is expected that a liberated Nigerian state would join other strong and well meaning nations of the world to put pressure on the USA to sign the protocol.

Furthermore, to halt the deplorable environmental, socioeconomic and political problems of the Niger Delta, the Nigerian state should declare a ten year national emergency plan to develop the Niger Delta. In doing this, the state should set aside about 20% of the present foreign reserve to fund the expected massive development projects in the Niger Delta. Apart from this, there should be the establishment of a Niger Delta Development and Investment Bank with an initial capital of about five billion dollars solely to support companies and businesses that may want to genuinely invest in that area. This would create more jobs and deplete the reserve army of militants engaged in criminal activities in the region. The federal government in partnership with other Niger Delta states should also initiate a Niger Delta Mega City project that would open up the cities through the massive construction of infrastructure that would encourage companies and businesses established in the Niger Delta to operate freely. It is also recommended that a string of satellite industrial and tourist cities should be built along the

East–West corridor to link all the states in the Niger Delta. To demonstrate sincerity and commitment to the development of the area, the plan and the earmarked funds should be placed under United Nations supervision like the oil for food project in Iraq or oil account for Chad to ensure transparency and international accountability.[39]

This chapter has provided copious evidence to conclude that the Nigerian state was illiberal from its inception. Rather than transforming the state, the new political elites inherited it, hence their inability to mediate contradictions and differences among the Nigerian people. It therefore had no choice but to become intolerant of opposition, relying on authoritarian and oppressive tactics. Those who were privilege to inherit the state from the colonial masters criminalized opposition so that they could continue to divert abundant resources that would otherwise have gone into developmental projects into their pockets and to purchase arms and ammunition to victimize, intimidate, eliminate and destroy voices of descent in Nigeria.

The chapter noted the failure of all the interventionist programs through the establishment of boards, commissions etc. and argued that they have failed because they were not set up to succeed in the first instance given the insensitive manner in which oil revenue accruable to states in the Niger Delta progressively decreased from one administration to the other. An inelegant role was played by the various legislations such as the Land Use Act, Petroleum Act, which permanently took the peoples' sources of livelihood away from them with very dire consequences for not only the people but also for their environment. This criminal expropriation created a situation where opulence and splendor stare extreme poverty and neglect in the face in the Niger Delta. It is also argued that the monopoly of technology, which the multinational corporations enjoy in the region, makes the state subservient to them and hence the multinational corporations' ability to violate extant environmental protection laws with impunity as well as subverting the state in oil exploration activities.

The point is that the various legislations enacted over the years by the Nigerian state are, to say the least, insensitive to the plight of the people and the environment in the Niger Delta region despite the fact that the area is a major source of oil revenue that accounts for about 96 percent of Nigeria's foreign exchange earnings. The people in host communities are not only neglected but compelled by the oil exploration activities of the state and the multinational corporations to live in shanties, abject poverty and squalor without the basic amenities of life such as drinkable water, hospitals, schools, roads and electricity.

Gas flaring and crude oil spillages are wrecking the worst havoc on the people and environment of the Niger Delta. Apart from its negative impact on agricultural output, because the people can no longer farm their lands or fish in the waters as a result of oil exploration, strange diseases hitherto unknown in the country are now manifesting in the region. Congenital abnormalities are rampant in the oil-bearing communities of

Eleme, Okirika, Umuechem, Oloibiri and Orashi region. For example, some children were born with one nostril, some with one and half and some with almost no nostril; others were born with mutilated lips. However, hospital sources revealed that these kids with these types of deformities do not live beyond six months.[40] All of these environmental problems make the Niger Delta area the poorest and the least developed in the Nigerian federation.

The chapter therefore concludes that unless the state is renegotiated along democratic lines, all interventionist efforts by the state may only be chasing the symptoms rather than the substance. The examples of the failure of the Niger Delta Development Authority, Niger Delta River Basin Development Authority, OMPADEC, NDDC and other interventionist programs are there to learn from. The point to note is that the illiberal nature of the state, the weakness of its institutions and the technical inadequacy of the regulatory agencies encourage unbridled environmental degradation in the region. Rather than strengthen these institutions, repression, exploitation and undue militarization of the entire region has been the order of the day. These weaknesses of the state and its apparatuses also makes it impossible for the state to compel oil multinational corporations to observe due diligence in oil exploration as well as implement environmental friendly policies in line with the extant laws.

Therefore, political inclusion, resource control and the principle of self-determination remains the pivot on which peace and progress can be restored to the Niger Delta because "the only way to secure the Niger Delta is to raise health, education and living standards, ensure free and fair elections, ameliorate conflicts over resources and broadly transform residents of the region into bona fide stakeholders who will benefit from oil revenue."[41] Unless the people are democratically empowered through well conducted, periodic, free and fair elections, which will give autonomy to the state in terms of its capacity to enforce environmental protection laws and development-oriented projects, all of the aforementioned ideals may be unattainable and the violent armed conflicts in the Niger Delta may go on unabated.

NOTES

1. For more on the recommendation of the Commission, see Sir Henry Willinks Commission Report in Nigeria, 1957.
2. John Locke (1960, 367) cited from Subrata Mukherjee and Sushila Ramaswamy, *A History of Political Thought: Plato to Marx* (New Delhi: Prentice-Hall of India, 2007), 20.
3. Mukherjee and Ramaswamy, 2007, 189.
4. Ibid., 201.
5. W. Doyle and Michael Kant, "Liberal Legacies and Foreign affairs," *Journal of Philosophy and Public Affairs* 12, no. 3 (Summer 1983), 206–207.
6. P. Dunleavy and B. O'Leary, *Theories of the State: The Politics of Liberal Democracy* (Basingstoke, England: Palgrave Macmillan, 1987), 1.

7. Kenneth Omeje, "Reappraising Contemporary Political Development in Nigeria," in *State-Society Relations in Nigeria: Democratic Consolidation, Conflicts and Reforms,* ed. Kenneth Omeje (London: Adonis & Abbey, 2007), 22.
8. Kenneth Omeje, 2007, 21.
9. Remi Anifowose, "State, Society and Nation" in *Element of Politics,* ed. Remi Anifowose and Francis Enemuo (Lagos, Nigeria: Malthouse Press Ltd., 1999), 98.
10. Eghosa Osaghae, *Crippled Giant: Nigeria since Independence* (Bloomington: Indiana University Press, 1998), 4.
11. Claude Ake, *Democracy and Development in Africa* (Ibadan, Nigeria: Spectrum Books Ltd., 1996), 2.
12. For a detailed discussion on the various dimensions of the legitimation crisis, See Adebayo Ninalowo, ed., *Crisis of Legitimation: The State and Vested Interest* (Lagos, Nigeria: Obaroh and Ogbinaka Publishers Ltd., 1999), 7–12.
13. Claude Ake, "The State in Contemporary Africa," in *Political Economy of Nigeria,* ed. Claude Ake (London and Lagos, Nigeria: Macmillan Press, 1985), 4–5.
14. See Kenneth Omeje, 2007, 21.
15. Akin Oyebode, "The Legal Framework of Oil Exploration in Nigeria—A Critique" in *Boiling Point,* ed. Wumi Raji et al. (Lagos, Nigeria: Committee for the Defense of Human Rights (CDHR), 2000), 55.
16. See Cyril I. Obi, "Globalization in Nigeria's Oil Industry: Implications for Local Politics" in *Nigeria's Struggle for Democracy and Good Governance,* ed. Adigun Agbaje et al. (Ibadan, Nigeria: Ibadan University Press, 2004), 44–45.
17. Oyebode, 2000, 58–59.
18. Oyebode, 2000, 58.
19. For more details on the Land Use Act, see Nigeria's Constitution, 1999.
20. Cyril Obi, 2004, 450.
21. Ibid.
22. Julius Ihonvbere, "Recipe for Perpetual Crises: The Nigerian State and the Niger Delta Question" in *Boiling Point,* ed. Wumi Raji et al., 90.
23. United Nations Development Program (UNDP) Report, 2006, 9.
24. Bisi Durotoye, "Geo-Environmental Constraints in the Development of Niger Delta Area of Nigeria" in *Environmental Problems of the Niger Delta,* ed. Akinjide Osuntokun (Lagos, Nigeria: Friedrich Ebert Foundation, 2000), 27.
25. Steve Okecha, "Environmental Problems of the Niger Delta and their Consequences" in *Environmental Problems of the Niger Delta,* ed. Akinjide Osuntokun, 52.
26. "End the Flares," *The Nation,* August 7, 2008.
27. Ibid.
28. For more detailed information about oil spills, see J. Shola Omotola, "From OMPADEC to NDDC: An Assessment of State Responses to Environmental Insecurities in the Niger Delta," *Africa Today* S4, 573.
29. Kenneth Omeje, "Oil Conflict in Nigeria: Contending Issues and Perspective of the Local Niger Delta People," *New Political Economy* 10, no. 3 (September, 2005).
30. Anayochukwu Agbo, "The Dying Delta," *The Tell Magazine* (Lagos, Nigeria: Tell Communications Limited), April 7, 2003, 59.
31. Ihonvbere, 2000, 183.
32. Examples of such brutalization include what is usually referred to as the "Umuechem Massacre" of 1990 where 80 people were killed while 490

houses were burnt down in Rivers state by anti-riots policeman. The police were sent there to quell protests over lack of electricity, pipe-borne water and other social amenities as well as for direct compensation for oil pollution of farmland and water resources. Umuechem at that time had 56 oil wells and operated two flow stations operated by Shell. There was also the "Odi Massacre" where a whole community was razed under the pretext that some youths killed a team of policemen sent to the village in 1999.

33. *Ablaze for Oil* (Abuja, Nigeria: Action Aid, 2008), 82.
34. Lucky Fiakpa, "A story of Rape," *The Tell Magazine* (Lagos, Nigeria: Tell Communication Limited), April 7, 2003, 53.
35. Fiakpa, April 7, 2003.
36. J. Shola Omotola, "From OMPADEC to NDDC," 573–589.
37. Mukherjee and Ramaswamy, 2007, 204.
38. Quote from Oronto Douglas, Deputy Director of Environmental Rights Action (ERA), in an interview with Dayo Aiyetan, "A Long-Drawn Struggle" *Tell Magazine* (Lagos, Nigeria: Tell Communication Ltd.), April 7, 2003, 45.
39. Chidi Amuta, "A Pox on this Summit," *The Nation Newspaper,* July 11, 2008, 40.
40. Anayochukwu Agbo, April 7, 2003.
41. Paul M. Lubeck et al., "Convergent Interests: U.S. Energy Security and the 'Securing' of Nigerian Democracy" (Ann Arbor, MI: International Policy Center, 2007), 5.

7 State and Governance of Multinational Corporations

Shell Petroleum Company in the Niger Delta

Samuel Ojo Oloruntoba

INTRODUCTION

The political arrangement represented by the Nigerian state has its root in the colonial strategy of capital accumulation, which dates back to the fusion of the Northern and Southern Protectorates in 1914. It is a contraption that one of the main actors in the early days of its evolution (the late Ahmadu Bello, the Sandauna of Sokoto) claimed was a consequence of the mistake of 1914. The Nigerian state that was granted independence on October 1, 1960, was a surrogate of colonial capitalism. It inherited from its foreign progenitor its socioeconomic structures, which were rooted in authoritarian, exploitative and predatory proclivities. It must also be emphasized that the Nigerian state functions as a mechanism for capital accumulation in the hands of domestic and foreign capital interests.[1]

The foundation of the Nigerian state is inherently flawed as it is structurally and externally induced to serve the interest of the local oligarchy and their international collaborators, who have been acting for their own interest both at the level of policy formulation and implementation. This preoccupation with crass material accumulation and manipulation of state institutions for pecuniary gains has weakened the capacity of the state to effectively enforce extant laws and regulate multinational corporations operating in the country. The import of the above to our discourse is that the Nigerian ruling elites have not demonstrated enough commitment to the effective mobilization of the human and material resources to galvanize a sustained process of socioeconomic and political development. The discovery of oil and the resultant dependency on this single product as at the main source of foreign exchange earnings has rendered the economy to be dangerously precarious and vulnerable to the vagaries of international commodity prices. As Obi notes, oil has become a defining factor and consideration for participating in the political process. He states, "oil provides the ample resources used in political transition . . . the closing of the gap between 'political power and oil power' thus renders oil a prime target of zero-sum politics, or the prize of political wars over the oil patrimony."[2] This has resulted in a situation where the multinational oil companies are left to operate with little or no control, which therefore predisposed them to all forms of subversion of joint-venture operational agreements, national laws and even international standards. For the past fifty years, the multinational oil corporations have operated with impunity in the wanton destruction of the natural habitat, the ecosystem and the degradation of the environment, the resultant effect of which is the pauperization of the indigenes of the Niger Delta. Olorode reinforces this point when he asserts that "unlike the experience with forestry, tinning and coal mining however, the oil industry not only assails the environment, it substantially excludes labour because of its high dependence on machine power."[3]

Further, as oil became the dominant source of foreign exchange and revenue earners (95% and 75%), respectively, it became the attracting force for participation in politics both by military ruling elites and their civilian counterparts. To achieve their aim, the political class and their collaborating bureaucrats perfected the art of effective manipulation and the abuse of the structures and processes of state power to accumulate oil surplus. Such practices, as Obi contends, are tendentiously marked by authoritarianism, corruption and parochialism, which not only breed instability and politics of exclusion but indirectly institutionalize violence as a matter of state policy.[4] The state largely relied on the use of force to silence the voices of dissent and to allow multinational oil companies to continue to pillage the oil resources of the Niger Delta at the expense of the welfare of the people. From the heroic declaration of the Independent Niger Delta People's Republic by Isaac Adaka Boro on February 24, 1966, to the struggles for recognition by the Movement for the Survival of Ogoni People under the leadership of Ken Saro-Wiwa, the Kaiama Declaration of December 11th, 1998, through to the nascent agitation for resource control by the Movement for the Emancipation of Niger Delta, among other groups, the Nigerian state has unleashed terror on the people to satisfy the interest of the tiny ruling elites and the multinational oil corporations.[5]

Falana graphically reinforces this view with the narration of the pathetic manner in which Adaka Boro and other leaders of the Delta Volunteer Service were charged with treason, tried, convicted and sentenced to death by the Port Harcourt High Court, which the Supreme Court also confirmed. The case of Ken Saro-Wiwa and members of his group was even more pitiable. He states that "Ken Saro-Wiwa and other MOSOP leaders were rounded up and charged with murder before the Special Military Tribunal, which convicted and sentenced them to death by hanging. The corpses of the Ogoni 9 were bathed in liters of acid after they had been hastily executed at the order of the late dictator, General Sani Abacha, on November 10, 1995."[6]

The overt and covert support of the state for multinational oil corporations emboldened them to continue their activities in utter disregard for the extant regulations governing the operations of oil companies in Nigeria. The state remains helpless and ineffective in enforcing these rules because of the greed, parochialism and the desire for the preservation of their economic status, which is made possible by rents and bribery from oil companies. Although the home governments of these multinational corporations have come under severe pressures from the civil society organizations and the nongovernmental organizations, no substantial efforts have been made to make the multinational corporations respect the codes of corporate governance and ensure environmental protection. This chapter is particularly concerned with how the ineffectiveness of the state has enabled Shell Petroleum (the biggest oil company in Nigeria) to serially degrade the environment of the Niger Delta through gas flaring, water and land pollutions and collude with the state apparatus of coercion to intimidate, suppress and in extreme cases emasculate individuals and organizations agitating for a fair share of the oil deal.[7]

In setting about this task, the chapter is divided into three parts: first, the conceptual and theoretical section dealing with the state, multinational corporations and governance. The second part deals with the governance challenge of the oil companies in the Niger Delta with focus on Shell BP. In the third and concluding section, the arguments are summed up and followed by an examination of the possibilities of restructuring the state in Nigeria through which the multinational companies can be brought under control and fulfill their corporate social responsibilities to the host communities while operating within the bounds of the legal framework in the country.

CONCEPTUAL AND THEORETICAL FRAMEWORKS

The Nigerian state like any other of its kind in Africa is a colonial contraption established to serve the purpose of capital accumulation in the interest of the developed world. A careful interrogation of the nature of the state in Nigeria can be situated within the context of Marxian analysis of class contradictions in any society. He conceived the ruling elites as a group of people who use the state as an instrument of domination of a social class or a coalition of social groups over the rest of the society. Although there is inequality in every society, with the powerful class exploiting the weak, this is more pronounced in the capitalist society, where commodification is the defining pattern of human relationships. To elaborate, commodification implies a form of existence in which maximization of profit is the main goal of entering any form of relationship and the main reason for carrying out any activity.[8]

The present condition of the state in Africa has its root in a historical trajectory, which manifests immanent successive characteristics. As Abdoulaye Bathily chronicled it in his "West African State in Historical Perspective," available historical evidence suggests the successive rise of five types of states in the subregion: these are (a) the primary state that was based on different social groups that performed complex lists of economic functions in society at a stage when primary economic activities such as farming, hunting and stick gathering among others were predominant; (b) a military merchant state of the trans-Saharan period of the eighth to sixteenth century where the Islamic religion was introduced as a substitute for the traditional religion, which led to militarization of the state and the resultant exploitation of the people by the dominant class; (c) the predatory state of the era of the Atlantic slave trade and colonial conquest in the seventeenth to nineteenth centuries with highly deleterious consequences arising from the export of millions of able bodied men from the subregion and integration into the modern world system; (d) the colonial state, which led to a dramatic structural alteration of the African social formation through the military defeat of the traditional rulers and the imposition of colonial power; and (e) the bureaucratic state of the post-colonial period where the "heirs" to the collaborators of the colonials days, the bureaucratic bourgeoisie, exert direct control on the state power.[9]

Although our analysis is essentially centered on the neo-colonial state, no doubt there is an unbreakable continuum in the historical trajectory of the state in this subregion that affects its relationship with the international capitalist system. Kola Olugbade in his "The State of the State of the Third World" observes that statehood not only represents a set of institutions but also a body of attitudes, practices and codes of behavior. As a matter of fact, he contends that the state in contemporary time fills the entire social space, with a pervasiveness that has become an inescapable reality.[10]

The theoretical underpinning of the state is as diffuse as the concept itself; hence, we have the sociological and ideological perspective, diffusionists and structuralists. Yet recent writings on the state have broadly drawn on two main streams of social and political thought—liberalism and Marxism.[11] According to Dyson, the state idea is a generalizing, integrating and legitimating concept that identifies the leading values of the political community with reference to which authority is to be exercised; emphasizes the distinctive character and unity of public power compared with civil society; focuses on the need for depersonalization of the exercise of the power; finds its embodiment in one or more institutions and one or more public purposes that thereby acquire a special ethic and prestige and an association with the public interest or general welfare; and produces a social-cultural awareness of (and sometimes of dissociation from) the unique and superior nature of the state itself.[12]

A number of other theorists have made intellectual contributions to the theory of the state as it affects the third world. Some of these include Gramsci, Althusser and Poulantzas. It is not our intention to go into the details of these theories; however, Gramsci's position provides a good understanding for our discourse. For instance, his emphasis is that ideas and ideology play a crucial role in the determination of economic structures and that a bourgeois society is not simply controlled by open force but that its mode of operation is through consent. To him, hegemony represents a subtle form of cultural domination and power is redefined in terms of intellectual hegemony. The masses are co-opted and quelled by means of ideational domination. The state is therefore not just a political or institutional apparatus that coercively dominates one class as evinced by Karl Marx; it is a vessel of intellectual dominance, which actually elicits a response from the masses.[13]

Even if this theory fails to capture the domestic dynamics of the internal contradictions of capital accumulation by the petit bourgeoisie, it is relevant in understanding the imposition of theories of development by the same international capitalist system through the Bretton Woods Institutions: the World Bank and the International Monetary Fund. These organizations posit that liberalization in terms of trade openness, foreign direct investment and privatization holds the key for sustainable development of a developing country like Nigeria. They also stridently advocate for the removal of restrictions on the operations of multinational corporations, reduction of taxes and overemphasize the role of foreign capital in the process of development. This point was emphasized by Terisa Turner's thesis on the triangular relationship between the state, foreign capital and the domestic bourgeoisie that has, in the process of accumulation, worked against the development of indigenous technology in the oil sector of Nigeria.[14]

In summing up the conceptual and theoretical positions of the Nigerian state within the context of the global capitalist system and its collaborative romance with multinational oil companies in their rapacious exploitation of the oil wealth in the Niger Delta, the submission of Daniel Omoweh is quite apt. He notes that in terms of character, the Nigerian state is not just capitalist but also transnational in the sense that it helps reproduce the global capitalist process of accumulation of which it is an integral part. We can then conclude that the state is a social public force whose objective is its monopoly and use of coercion not to police and guide the society rather impartially because it has to meet the conflicting interests of the institutional groups that make it a reality. Instead, it is a specific public force dominating the larger society through the system of institutional or class mechanisms to cater to its selfish, political, economic and social interest and those of its local fronts (the comprador bourgeoisie) and foreign capital in all sectors of the Nigerian economy including the energy sector.[15]

The current situation in the Niger Delta and the role of the multinational oil companies in exploitation and expropriation of oil resources in this region is in synchronization with Terrisa Turner's and Omoweh's points above that the alliance between the state, the multinational oil companies and the indigenous comprador has not led to technological development of the area. Beyond this, however, is the pauperization of the people through the destruction of the ecosystem and by implication their means of livelihood as a fishing and farming society; the militarization of the region by the agents of the state; and the suppression of protest and the resultant emasculation of the leaders of the various movements seeking fair balance in the sharing of oil revenue. Examples are the hanging of Ken Saro-Wiwa and eight other Ogoni indigenes in 1995 and, until recently, the detention of leaders of the Movement for the Emancipation of Niger Delta (MEND), Mr. Henry Okah and Asari Dokubo of the Niger Delta Volunteer Force.

MULTINATIONAL CORPORATIONS

Multinational corporations (MNC) are privately owned business enterprises incorporated under the laws of a particular country but that engage in such activities as mining, industrial, commercial and agricultural production in a country or a number of countries but certainly other than their nationality.[16] In a more elaborate form, Raymond Vermon provides another description and distinguishing features of a multinational corporation. A multinational corporation is a parent company that controls a large cluster of corporations of various nationalities. The corporations that make up each cluster appear to have access to a common pool of human and financial resources and seem responsive to elements of a common strategy. Size is important as well; a cluster of this sort with less than N100 million in sales rarely merits such attention. Moreover, the nature of the group's activities outside its home country is relevant; mere exporters, even exporters with well established subsidiary sales abroad, are unlikely to draw much attention, and mere licensers of technology are just as rarely mentioned.[17]

Finally, the enterprises involved generally have a certain amount of geographical spread; a patent with a stake in only a country or two outside its home base is not often found on the list of multinational enterprises. Implicit in this definition are four characteristics of the multinational corporation: size, central control, geographic spread and distribution. Essentially, a multinational corporation as Joan Spero opined "is a firm with foreign subsidiaries which extends the production and marketing of the firm beyond the boundaries of any one country. They are firms which have sent abroad a package of capital technology, managerial talent and marketing skills to carry out production in foreign countries.[18]

Adejumobi agrees with Akinsanya that the two characteristic features of MNCs are that decision making is centralized and the subsidiaries operate under the discipline and framework of a common global strategy and common global control from the headquarters. Second, MNCs normally have two levels of operations: the first is in the home country where the headquarters are located and decisions are taken and the second level is the host countries where subsidiaries are located.[19]

In the aftermath of the Second World War, two main strands of argument permeated the literature on the purpose and the relevance of MNCs to the development of third-world countries. The orthodox theoretical perspective sees multinational corporations as making possible the use of total world resources with the maximum of efficiency and the minimum of waste and whose presence is vital for the rapid growth and development of all economies, especially the less developed countries.[20] Because MNCs pursue a global scope in areas of research, marketing and production, it is argued that they represent the transcendence of politics in the interest of world peace and development. The radical perspective, which has a Marxian orientation, sees the MNCs as the most important aspect of the imperialist penetration and exploitation of third-world countries by heightening their dependency and deepening the process of dependent development. According to this school of thought, the MNC is an instrument of the international struggle supplanting the bourgeoisie nation state precisely because it is a more efficient means by which the capitalist economies can dominate and exploit the less developed countries.[21]

As Akinsanya further elaborates, the Marxist argument against MNCs is essentially premised on three grounds. First, because of internal contradictions in a capitalistic system, large business enterprises (i.e., the MNCs) are compelled to establish subsidiaries overseas. This doctrine of institutional necessity for direct foreign investment, which is very fundamental to the Marxist theory of imperialism, tallies with Hobson's liberal theory of imperialism, which states that the search for foreign direct investment outlets is a function of the law of the falling rule of profit, which is due in turn to the accumulation of surplus capital in matured capitalist economies. Second, the foreign policy of the United States, Europe, Japan and of late China is a reflection of the expansionist interests of the dominant bourgeoisie class. The third and most significant primary goal of this dominant (bourgeoisie) class is to eliminate all obstacles (e.g., the nation state) to corporate expansionism and to increase profits at the expense of less developed countries.[22]

We share the views expressed above, as it is evident that since 1956 when the Shell Petroleum Development Company started operations in Nigeria, the company forged alliances with the indigenous capitalists to exploit the oil resources in the Niger Delta to the detriment of the welfare of the people in the region. The corporate social responsibility of the company failed to address environmental hazards that continue to pose threats to the livelihood of the people of the region. For instance, despite official pronouncements to the contrary, the company persists in gas flaring, which portends a huge economic loss as well as endangers the ozone layer with its far-reaching health implications on the lives of the people.

It is incontrovertible that environmental problems caused by the activities of multinational corporations include freshwater pollution, air pollution, chemical pollution, soil and land pollution and degradation, industrial pollution, oil and gas pollution, urbanization and unsustainable agricultural practices such as fish farming, malnutrition and disease, freshwater shortages, habitat and ecological systems, etc. All of these have combined to worsen the plight of the indigenes of the region. Rather than the state enforcing the necessary rules and forcing the multinational corporations to comply, it has resorted to the use of brute force to kill, maim and subjugate the indigenes of the region. There is however, a middle of the road school that takes the position that within a modified framework and working with an agreed set of guidelines and rules, the multinational corporations, through the ownership and control of technology, capital and marketing skills, can make a unique contribution to the development effort (sic of the third-world countries).[23]

Even if it is true that multinational companies can foster economic development due to the ease with which they have access to capital and the high quality of their staff, it is the contention of this chapter that this can only happen if there is effective control from the state. In other words, the state must be willing and capable of enforcing the regulatory frameworks under which the multinational companies should operate in the country. It is the failure of the Nigerian State to effectively control Shell and other oil companies in the country that has provided these companies with the undue latitude that they have in failing to keep with the environmental standards and performing their corporate social responsibilities to the host communities of the Niger Delta. It is the capacity of the state to effectively monitor and control the activities of multinational companies to act for the interests of all stakeholders and not just their shareholders, most of whom are resident abroad.

GOVERNANCE

The term governance has become almost synonymous with civil or political governance. Governance itself is a multidimensional concept used to describe appropriate control and the enforcement of standards either in political, economic or corporate governance. As Otobo argues, the relationship among political, economic and corporate governance is analogous to a series of concentric circles in which political governance forms the outside circle, followed by the economic governance circle, with corporate governance at the centre.[24] Figure 7.1 illustrates the concept.

Political governance sets the orientation of the economy, for example, whether a country will be a market economy, mixed economy or centrally planned economy. Political governance supplies the institutional infrastructure for economic governance and by institutional infrastructure it is meant both the rules of the game and the organizational structures. In turn, economic governance provides the context in which corporate governance is practiced by providing the laws under which corporations are established, the regulatory framework for the conduct of corporate affairs and the macro-economic framework.

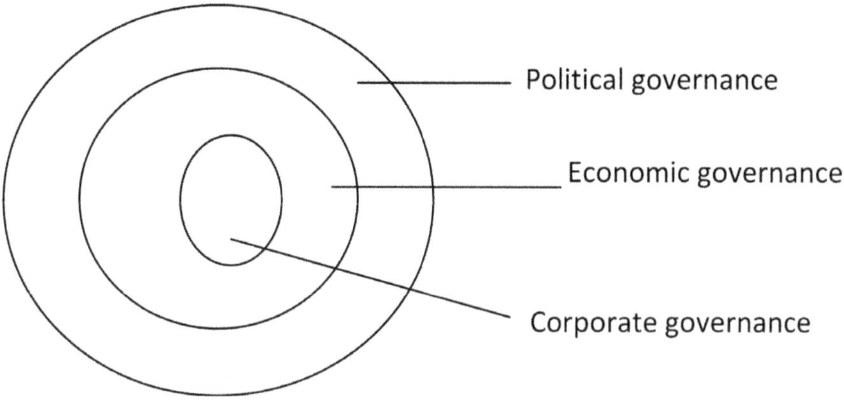

Figure 7.1 Diagram of the interaction of spheres of government in Nigeria.

The bankruptcy of companies that were once successfully managed in the United States of America and Britain, like the Enron Corporation, Pamellat, Amazon and of recent Lehman Brothers among others, has generated interest in the governance of companies across the world. In particular, there are external influences that reinforce the layer of external regulations on corporate governance. Otobo further avers that external regulations means the body of codes, norms or laws imposed on the corporation either by the state or through international agreements. These laws, codes or norms cover such areas as antitrust, securities and environmental laws.[25]

Parkinson defines corporate governance as the process of supervision and control (of governing) intended to ensure that the company management acts in accordance with the interests of the shareholders.[26] There are two main models of corporate governance, the stakeholders' and the shareholders' models. In the stakeholders' model, corporate policies and strategies take into account not only the interest of those who invest in (shareholders), work for (managers) and direct the affairs of the corporation (the board of directors), but also the employees, suppliers, the community and the workers union. By contrast, in the shareholders model, the prime focus is on earning returns to the shareholders, the board and the management.

THE GOVERNANCE CHALLENGE OF MULTINATIONAL CORPORATIONS IN NIGERIA

We have established that the character of the Nigerian state weakened its ability to enforce laws and regulations governing the operations of multinational corporations, most especially in the oil rich Niger Delta. In this section, we address the relevant laws that guide oil operations in Nigeria. The essence of this is to show that the excesses of multinational oil companies in terms of environmental degradation and pauperization of the people of Niger Delta is not on account of absence of relevant laws and standards to regulate their activities. Rather, it is a manifestation of crass ineptitude on the part of the Nigerian State to enforce these laws, which in itself is a demonstration of willful conspiracy of the ruling elites with the multinational oil corporations to expropriate oil and gas resources for personal and class profits, to the detriment of the interest of the people.

The legal framework that governs the operation of multinational oil companies in Nigeria was set out in the Minerals Act of 1958. This Act intended to amend and consolidate all the existing legislations relating to mines and minerals and it vested ownership of mineral resources including petroleum in the federal government. More importantly, both the independence and republican constitutions empowered the federal government, in an exclusive capacity, to pass legislation on mines and minerals including oil fields, oil mining, geological surveys and natural gas.[27]

In the First Republic, the constitution of Nigeria vested ownership of natural resources on the respective regions and in consideration of this, a region is entitled to 50 percent of mining royalties and rents on minerals and other resources derived from it. However, the military incursion into Nigerian politics in 1966 and the promulgation of the Petroleum Decree of 1969 vested the entire ownership and control of all petroleum in, under or on any lands in the state, in the federal government. Emerging from the above is the fact that Nigerian petroleum law includes three types of licenses: the oil exploration license, the Oil Prospecting License and the Oil Mining Lease.

The Oil Exploration License (OEL) as envisaged under the Petroleum Decree merely enables the licensee, originally limited to citizens of Nigeria and companies incorporated under Nigerian law, to conduct aerial and surface geological survey excluding below 300 feet. The oil prospecting license (OPL), like the oil exploration license, is a preproduction license that enables the licensee to undertake more detailed and expensive investigation of the allotted area with a view to ascertaining its viability in terms of petroleum production as suggested by the initial explorations of the holders of the OPL. The holders are empowered to conduct drilling operations to any depth, and if they discover petroleum in the area as prescribed under the license, such holders are given the right to drill and export it. However, the life of an OPL shall not exceed five years in the case of land and territorial sea or seven years for the continental shelf.

The Oil Mining Lease (OML) may be granted to the holder of an oil prospecting lease who is deserving of the same. The holder of an OML is required to relinquish one half of the area of the lease ten years after the grant and the lessee is prohibited from assigning his lease without the consent of the Minister of Petroleum Resources and not after the payment of the prescribed fees. The law also provides among other things that the lessee must exercise due diligence by taking all practicable means. However, grantees under the repealed Mineral Oils Act conferred on them all or any petroleum lying within or under the leased areas described in the lease for a period of thirty years with an option to renew for a second thirty-year period for land areas and forty years, in the first instance, and with a right of renewal for another forty years in the case of the continental shelf. Under the saving provisions of the Petroleum Decree, all leases granted under the repealed legislation were validated for not more than twenty years after the promulgation of the decree.

It is instructive that the holder of an OML is obliged to respect the rights of any person or persons in lawful occupation of land in his area of operation by entering into negotiation with such for payment of fair and adequate compensation for the disturbance of their surface rights. In addition, the lessee carrying out his activities should ensure that he does not "deprive any land, village, houses or watering places for cattle of a reasonable supply of water or interfere with any rights enjoyed by any person." Accordingly, the lessee must be ready to pay adequate compensation for things such as economic trees, structures affixed to land, objects etc. Besides, he is debarred from tampering with sacred lands, burial grounds, public installations, markets, etc. within a specific distance from his area of operation. In sum, the lessee must be mindful of the requirements of conservation and sustainable development all of which have become concretized in good oil field practices.

Accordingly, the lessee must exercise due diligence by taking all practicable means to control the flow and to prevent the escape of petroleum discovered in, or obtained from, the relevant areas, to prevent damage to adjoining petroleum bearing strata except for the purpose of secondary recovery as authorized by the Directorate of Petroleum Resources (DPR) to prevent the entrance of water through boreholes and wells to petroleum-bearing strata; to prevent the escape of petroleum into any water, well, spring, stream, river, lake, reservoir, estuary or harbor; and to cause as little damage as possible to the surface of the relevant area and to the trees, crops, building, structures and other property thereon.[28]

However, these legal provisions have been routinely flouted by the multinational oil companies owing to the fact that they receive protection from the state. Again, the structure of the Federation and the abrogation of the previous practice, which operated on the principle of derivation, worsened the problem. Consequently, the Niger Delta region is denied its rightful due on the oil and gas exploited therein from the federation. Cosmetic approaches like the establishment of Oil Minerals Producing Area Development Commission (OMPADEC), Niger Delta Development Commission and of recent, the Ministry of Niger Delta is a mere distraction. As previous experiences have shown, these organizations only created opportunities for corrupt politicians to further enrich themselves at the expense of the majority of the citizens who have been disempowered and pauperized.

A corollary to the defective federal system of government, which has made it impossible for the oil-producing areas to derive maximum benefit from the product, is the nature of the Nigerian political arrangement and practice where the major ethnic groups, the Yoruba of the South West, the Hausa Fulani of the North and the Igbo of the East have dominated the political space to the exclusion of the minority areas since independence in 1960. Because of the political advantage that these groups have over the minorities, policies regarding resource allocation have been done disproportionately to favor the majority groups. This was worsened by the many years of military dictatorship, where no pretension was made to any idea of true federalism.

Shell, a British company, was the first to start oil exploration in Nigeria. In 1956 oil was discovered in Oloibiri in present day Bayelsa state. However, commercial exploration and exploitation started effectively in 1958. As the first among equals, this company controls the largest oil field and by implication the highest volume of oil production and exports in Nigeria. Over the past four decades, the company systematically worked in connivance with the rapacious stealing political and business elites in Nigeria to circumvent virtually all available regulations that govern oil exploration and production in the Niger Delta. This company operates with little or no concern for the environment. As the company with the largest oil field, it represents, in a graphic manner, the destructive activities of multinational oil companies in the Niger Delta. As noted above, they participate in incessant oil spillage, gas flaring and the emission of hydrocarbon materials into the air. These result into acid rain, which rendered roofing sheets unusable after two years as against the normal life span of ten years in that region, noise and air pollution, which have deleterious health hazards for the inhabitants of the area. All of these combined have rendered the lives of the people in this region to be precariously dangerous, unproductive, nasty and unfulfilled. Usually, the oil companies attribute oil spills to sabotage by residents of the oil producing states. Curiously, in a clear case of conspiracy with the ruling elites who make laws for the country, under the Nigerian environmental laws, companies are not obliged to clean up or compensate for the effects of spills caused by sabotage.

Another instance of the weakness or the failure of the Nigerian state to enforce extant laws and ensure compliance with standard corporate governance is in the area of gas flaring. Even though the government has set several deadlines in which the multinational oil companies are supposed to stop gas flaring, the companies led by Shell have consistently given one excuse or the other why they could not stop flaring gases, which, as we noted earlier in this work, constitute a huge economic loss and serious environmental hazards for the people of the Niger Delta. The latest in the series of such breaches was the December 2008 deadline. As the deadline approached, the oil companies contemptuously gave notice to the government that it was not possible for them to meet the deadline. The penalty the state claims that it imposed on the companies for failing to comply was a paltry sum, which the companies might not even pay.[29]

As expected, there have been internal and external responses to this injustice. Yet Shell has been leveraging its alliance with the ruling elites to subdue every voice of dissent at the local level while employing sophisticated propaganda and public relations at the international level to deceive human rights activists and stakeholders in the campaign for justice and environmental sustainability in the Niger Delta region. For instance, the agitation by Ken Saro-Wiwa and his group, The Movement for the Survail of Ogoni People (MOSOP), against the plundering of the environment in Ogoni land led to the intervention of the state, which eventually led to the de-capitulation of Ken Saro-Wiwa and the eight members of his group through hanging. The current occupation of the Niger Delta by various security outfits like the Joint Task Force (JTF) and Operation Restore Hope are indicators of the sympathy of the state with the oil companies.

The company has also been playing the various communities in the Niger Delta against one another in their quest for unchallenged operations and exploitation of the resources in the region. This policy of divide and rule is orchestrated through the sue of rents payment and other patronage like award of contracts, most especially to politicians, who in turn recruit jobless and frustrated youths with whom they unleash terror on those perceived to be taking hard line positions.[30]

The Nigerian Extractive Transparency Initiative, which was set up to ensure accountability in the inflow of oil resources in the region has recently lamented that there is a gross discrepancy in the accounts of state institutions charged with managing the revenue inflow of these resources, first among themselves and secondly between the multinational oil companies under the so-called Joint-Venture Agreement. Such institutions include the Central Bank of Nigeria, the Nigerian National Petroleum Corporation and the Federal Ministry of Finance. Such willful negligence on a simple matter like tracking the volume of oil exploitation and export further corroborates the argument of this paper that the Nigerian state has been a willing accomplice with the multinational oil companies in their rapacious expropriation of oil for profit maximization at the expense of the local indigenes of the Niger Delta. For instance, as Okonta explains, "not a single public health survey has been conducted in the Niger Delta to fully study the impact of over four decades of oil production on the people. How much does the company (Shell) earn from millions of barrels of crude oil of the Niger Delta everyday? No one knows, neither Nigeria's Auditor-General nor Minister of Finance."[31]

Also Nigeria is the eighth largest oil exporter in the world. The state has, through collaboration with the oil companies, failed to ensure that the three refineries in the country work to their full capacities. Rather, in a somewhat strange and economic twist of logic, the country exports about 90% of its crude oil and imports almost the same percentage in refined petroleum products, thereby denying the oil-producing communities and Nigeria as a whole the benefits that could have accrued from refining crude oil in the country, in terms of higher revenue flow, creation of employment and establishment of allied small- and medium-scale industries in the petrochemical industry subsector. During a recent visit to Ghana in the course of this research, we discovered that whereas Shell operates filling stations in this neighboring country, the company does not have a single refinery or filling station in Nigeria, yet the state looks on as if nothing is amiss.

At the international level, there have been responses from nongovernmental organizations like Human Rights Watch, Environmental Rights Action etc. from the advanced companies. These organizations have successfully brought into the international public space the obnoxious activities of these organizations, especially Shell Corporation.Yet, the company prefers to engage in arm twisting and rhetoric rather than provide the needed empowerment for the indigenes of the host community.

CONCLUSION

We have looked at the role of the Nigerian state in the failure of the Multinational Corporations in Nigeria to meet the aspirations of the people of the Niger Delta. We have also established that the nature and character of the state and its incorporation into the international capitalist agenda have made it difficult, if not impossible, to enforce the extant laws regulating the activities of oil companies in the country. Over the years, multinational oil companies acted in ways destructive to the environment and by extension, the livelihood of the host communities in the Niger Delta region.

This situation engendered widespread condemnation and protest with wide ranging implications for the security of lives and property in the area. The resultant effect has been the subjugation of the various groups like the Movement for the Emancipation of Niger Delta (MEND), the Niger Delta Volunteer Force and so on. Other than the use of force, the Nigerian State also employed the use of carrot-and-stick tactics, a combination of appeasement, subterfuge, infiltration and dilution through class alliances to weaken the resolve of the individuals and organizations that are committed to getting Shell to fulfill its corporate social responsibility to the people. For instance, during the Obasanjo administration from 2003 to 2007, the leader of the Niger Delta Volunteer Force, Asari Mujahdeen Dokubo, was flown in a Presidential plane from Port Harcourt to Abuja, ostensibly to negotiate terms of peace with him. However, a few weeks after that, he was arrested and held incommunicado for plotting to overthrow the government. He was also charged for committing treason.

We have also noted that despite the high-handedness of the state and the complicity of the oil companies, particularly the Shell Corporation, violence in this region has continued unabated by kidnapping expatriates, and the blowing out of oil flow stations both onshore and offshore have reached a frightening dimension. All these have a deleterious effect on the economy of Nigeria, which is essentially monolithic. We argue that the problem will continue as long as the state maintains a cosmetic approach to resolving the intricate issues at hand.

It is our conclusion that to find a lasting solution to this problem the Nigerian Federation must be restructured. The many years of military rule led to unitary federalism, which in itself is a contradiction in terms. It is therefore pertinent that the issue of resource control should be given serious consideration. This will ensure that, as it obtained in the First Republican Constitution, 50% of the revenue from the resources procured from a unit of the federation will be reinvested back into the development of the unit, whereas the remaining 50% will go to the Federal Government Account.

Besides, there is need for more citizens' actions against the rapacious and parasitic ruling elites to make the state to be more responsive to the yearnings of the people. It is also imperative that the citizens should be involved in the process of enacting various laws that affect the exploitation of oil and gas in the Niger Delta region. This will ensure inclusiveness of the view of the people and guarantee that their interests are taken care of. A strong state is equally necessary to enforce the rules governing the operations of the multinational oil companies in the country. This is against the backdrop of the fact that these companies are capitalistic in their orientation and they will do everything possible, regardless of the ethical consequences, to maximize profits. As obtainable in other countries, the state in Nigeria should invoke all necessary codes of corporate governance from both domestic and international levels to elicit compliance to environmental standards in the locations of oil operations.

Finally, as violence may lead to a loss of sympathy for organizations championing the cause of the people in the Niger Delta, these organizations should adopt other instruments like propaganda and publicity in the print and electronic media both locally and internationally to draw further attention to their plights. Judicial redress should also be sought. Even if it is assumed that the Nigerian Judiciary is an integral part of the predatory and oppressive capitalist state, individuals and organizations in this region can seek for redress in regional courts like that of the Economic Community of West African States (ECOWAS), the African Union or the International Court of Justice. Also, international nongovernmental organizations such as Amnesty International, Green Peace and Friends of the Earth should continue to mount pressure on the home governments of the multinational corporations to enforce extant laws on the protection of environments where they operate. They should also compel them to perform their social responsibilities to the host communities in terms of employments, payment of royalties and provision of scholarships to students from these areas.

NOTES

1. Claude Ake, "The State in Contemporary Africa," in *Political Economy of Nigeria,* ed. Claude Ake (London: Longman, 1985). See also Julius Ihonvbere, "Constitutionalism and the National Question in Nigeria," in *Nigeria's Struggle for Democracy and Good Governance, A festschriff for Oyeleye Oyediran,* ed. A. Agbaje Adigun, Larry Diamond, and Ebere Onwudiwe (Ibadan, Nigeria: Ibadan University Press, 2004).
2. Cyril Obi, "Oil and Transition Politics," in *Transition Politics in Nigeria 1970–1999,* ed. Browne Onuoha and M. M. Fadakinte (Lagos, Nigeria: Malthouse Press Ltd., 2002), 97. For a more detailed account of the acquiescence of the Nigerian state to the excesses of multinational oil companies in the Niger Delta, see Innocent Ojeifa, "The Nigerian State and the Environmental Question," in *Environmental Problems of the Niger Delta,* ed. Akinjide Osuntokun (Lagos, Nigeria: Friedrich Ebert Foundation, 2000). The point was also made more poignantly by Ihonvbere that although oil is produced in the minority states, the majority ethnic groups dominated its recycling and allocation. Julius Ihonvbere, "The Recipe for Perpetual Crises: The Nigerian State and the Niger Delta Question," in *Boiling Point, a CDHR Publication on the Crises in the Oil Producing Communities,* ed. Wumi Raji, Ayodele Ale, and Eni Akinsola (Lagos, Nigeria: CDHR, 2000).
3. O. Olorode, "NGOs, Military Rule and Popular Empowerment," in *Nigeria: Non-Governmental Organizations and Democracy,* ed. O. Olorode et al. (Lagos, Nigeria: CDHR, 1997), 13.
4. Cyril Obi, 2002, 98.
5. Wumi Raji, *Boiling Point.*
6. Femi Falana, "Crises in the Oil Producing Communities in Nigeria," A Welcome Address at the Conference organized by the Committee for the Defence of Human Rights (Lagos, Nigeria: CDHR, 2002), 2–5. Also see Frynas Jedrzri, "Corporate and State Responses to Anti-Oil Protests in the Niger Delta," *African Affairs* 100 (2001), 27–54.
7. Olayiwola Owolabi and Iwebunoh Okwechime, "Oil and Security in Nigeria: The Niger Delta Crisis Africa Development," *Council for the Development of Social Science Research in Africa* 32, no. 1 (2007): 1–40.
8. Samir Amin, "Foreword," in *Globalization, Inequalities and the Commodification of Life and Well-Being,* ed. Mammo Muchie and Li Xing (London: Adonis & Abbey Publishers Ltd., 2006).
9. Abdoulaye Bathily, *The West African State in Historical Perspective* (Dakar, Senegal: CODESSRIA, 1999).
10. Kola Olugbade, "State of the State of the Third World," *African Journal of Political Economy* 4, no. 1 (1989): 10–25.
11. Brian Head, "Recent Theories of the State," *Politics* 19, no. 1 (1984).
12. Kenneth Dyson, *The State Tradition in Western Europe* (New York: Oxford University, 1980).
13. Quintin Hoare and Georgewell Smith, eds., *The Prison Note Book of Anthonio Gramsci* (New York: International Publishers, 2005), 3–14.
14. Terisa Turner, "Commercial Capitalism and the 1975 Coup," in *Soldiers and Oil: The Political Transformation of Nigeria,* ed. K. Panter-Brick (London: Frank Cass, 1978).
15. D. Omoweh, "Oil Exploration in Nigeria—A Theoretical Overview," in Raji Wumi, *Boiling Point.*
16. Raymond Vernon, *Sovereignty at Bay: The Multinational Spread of U.S. Enterprises* (New York: Basic Books, 1971).

17. Adeoye Akinsanya, *Transnational Corporation and Economic Nationalism in the Third World* (Ilorin, Nigeria: Unilorin Press, 1984).
18. Joan Spero, *The Politics of International Economic Relations* (Belmont, CA: Wadsworth Publishing Company Ltd., 2003).
19. Said Adejumobi, "The Structural Adjustment Program and Multinational Corporations in Nigeria: Impacts and Implications," in *African Journal of International Affairs and Development* 2, no. 1 (1997).
20. David Colman and Nixon Fredrick, *Economics of Change in Less Developed Countries* (Oxford, England: Phillip Allan, 1978).
21. John Hobson, *Imperialism, A Study* (London: George Allen Unwin, 1938). His accumulation theory centers on the accumulation of surplus capital during the Industrial Revolution. This theory linked the problem of shrinking continental markets driving European capital overseas to an inequitable distribution of wealth in industrial Europe. This prompted imperialistic adventures under which colonies became a cheap source of raw material supplies. The theory further posits that the wages of workers did not represent enough purchasing power to absorb the vast amount of capital accumulated during the industrial revolution. Hobson's accumulation theory suggests that capitalism suffered from consumption due to the rise of monopoly capitalism and the concentration of wealth in fewer hands, which apparently gave rise to a misdistribution of purchasing power. He concluded his theory by saying that finance capital, which is an inherent feature of imperialism, was manipulating events for its own profit but often against broader national interests.
22. Hobson, 200.
23. David Colman and Frederick Nixon, 217.
24. Ejeviome Otobo, "External Influences on Corporate Governance," in *African Development and Governance Strategies in the 21st Century: Looking Back to Move Forward,* ed. Bade Onimode et al. (London/New York: Zed Books, 2004).
25. Ejeviome Otobo, "Institutions and Economic Governance for a Market Economy: Pathways for Africa," *Development Policy Management Network (DPMN) Bulletin* 6 (November 1 1999).
26. James Parkinson, *Corporate Power and Responsibility: Issues in the Theory of Company Law* (Oxford, England: Clarendon Press, 1994).
27. Akin Oyebode, "The Legal Framework of Oil Exploration in Nigeria" in *Boiling Point,* 53–64. For more on this, see Victor Ukaogo, "Transnational Business Ethics, Government Policies and the Crises of Pollution and Underdevelopment in the Niger-Delta," in *Environmental Problems of the Niger Delta,* ed. Akinjide Osuntokun (Lagos, Nigeria: Friedrich Ebert Foundation, 2000).
28. Oyebode, 57.
29. Ike Okonta, *Litmus Test: Obasanjo and the Niger Delta* (ERAction, January–March, 1999). For more practical examples of environmental degradation of the Niger-Delta, see David Aworawo, "The Impact of Environmental Degradation of the Rural Economy," in *Environmental Problems of the Niger Delta,* ed. Akinjide Osuntokun (Lagos, Nigeria: Friedrich Ebert Foundation, 2000).
30. Hardens Monique et al., eds., *Lessons Not Learned: The Other Shell Report* (Inkworks Press, 2004). www.foe.co.uk/resource/reports/lessons-not-learned.pdf. Accessed 01/22/09. Also see Adefemi Isumonah, "Oil Bearing Minorities' Struggle in Nigeria: Towards an Alternative Constitutional Framework," *Proliteria* 20, no. 2 (2001): 5–22. See "Ken Saro Wiwa," *Review of African Political Economy* 22, no. 66: 471–480.
31. Okonta, *Litmus Test.*

8 "Big Motor and Dirty House"

Ken Saro-Wiwa's Fight with Royal Dutch Shell and Nigeria

Michael Sharp

> "But is it a good thing to fight?" The short man was asking as he chopped *ngwangwo* from a plate.
>
> "I like to fight. Yes. It is a good thing to fight. If somebody take your thing by force, if 'e want by force you to do something wey you no like to do, then you fight am."
>
> Ken Saro-Wiwa, *Sozaboy* (1985)

The man whose name is now synonymous with human and environmental activism has continued to grow in importance as a world-wide symbol of nonviolent protest against the greed of Big Oil and corrupt government. The life and work of the Ogoni writer Ken Saro-Wiwa continues to arouse the passions of activists and writers alike in Africa and around the globe. The renaming of a street after him in Dublin, Ireland, in 2008 parallels a similar proposal to name a street after Saro-Wiwa in Abuja, the Nigerian capital. The Nobel laureate, Wole Soyinka, used this opportunity to call for the renaming of streets honoring former dictators. Just as the eminent Caribbean poet Kamau Brathwaite has warned of a coming "Age of Dis Dis.tress,"[1] Saro-Wiwa lives in the public consciousness where his writings remind us that life and land are sacred and must be lived and worked by "good faith, mercy, fair-dealing, honesty, and truthfulness."[2] If not, then life and land will fall deeper into what Brathwaite calls "Dis. Tress Dis. Pair & Dis. Respect. Dis. Trust. Dis. Rupt. Dis. truction."[3] As Saro-Wiwa went defiantly to the scaffold, he reportedly shouted, "Lord, receive my soul, but the struggle continues."[4]

When Saro-Wiwa took on the "Big motor" of Royal Dutch Shell and the "dirty house"[5] of Nigerian politics in the name of the Ogonis, whose lands and rivers in the Niger Delta had been devastated and polluted by decades of crude oil extraction, he knew that the nonviolent Movement for the Survival of the Ogoni People (MOSOP) he had helped to launch would meet with fierce resistance. This resistance would come not only from what Soyinka has labeled the "moneyed thuggery" of the multinational oil companies[6] but also from a "poli-thug state" run by a succession of military dictatorships.[7]

Like Soyinka, Saro-Wiwa accepted "Nigeria as a duty;"[8] but both men also understood that the country "cannot be the burden of the writer and the intellectual alone." Unlike Soyinka, however, Saro-Wiwa believed almost messianically that "God had sent him"[9] to speak for the unrepresented Ogoni whose livelihood had been disrupted by speculative drilling and "the harsh nature of domestic Nigerian colonialism."[10] Whether Ogoniland was menaced by Major General Ibrahim Babangida's "bloodhounds"[11] or cut to pieces, in Soyinka's words, by "the Nigerian butcher,"[12] Major General Sani Abacha, Saro-Wiwa made it his life's work to grapple with what he called the "Satanic Octopus,"[13] that tentacular collusion between Royal Dutch Shell in the Niger Delta and Aso Rock, the seat of periodic military governments, in Abuja.

Saro-Wiwa believed in the Romantic concept of the burden of the poet. He understood the poetic (and prophetic) role of the Ancient Mariner in Samuel Taylor Coleridge's eponymous poem who importunes the reader with a vision of what he has seen. Similarly, Saro-Wiwa identified with William Wordsworth's belief that he was "a chosen son."[14] He felt instinctively that "The failure to use the creative imagination has reduced Africans to the status of mimic men and consumers of others' imagination."[15] Like Wordsworth, whom he had read while a student at the University of Ibadan, Saro-Wiwa felt he had "holy powers/And faculties."[16] Whereas he was once called "Saint Ken–Violent Hero of the Gullible,"[17] encouraging his enemies to disparagingly compare him to what Eric Hobsbawm has called 'the primitive rebel,'[18] Saro-Wiwa continued to inspire Ogoniland and its people with both an infectious optimism and a bitter pessimism. This is reminiscent of the poet Percy Shelley who, in "The Mask of Anarchy" (1819), imagined peaceful demonstrators at a political rally standing shoulder to shoulder as they are charged by an incensed cavalry. Shelley, a self-confessed idealist, nevertheless, computed the mathematical irony of the final line: "Ye are many–they are few."[19] Neither "Shellspeak"[20] nor the threats of a brutal dictatorship, however, were able to take away from what Mary Harvan has sensitively called Saro-Wiwa's "nurturing role," which "shifted from public leadership to spiritual inspiration."[21]

Declaring that "A writer is his cause,"[22] Saro-Wiwa understood Nigeria was "a sleeping elephant" and that to wake "the big, bad beast from its stupor, a sledgehammer [was] needed."[23] He also knew that Ogoniland and its people would "suffer as the grass in a fight between two elephants."[24] Literature would do the waking, and Saro-Wiwa's "recreating the Ogoni people", would be also "for the rest of Africa in the course of time."[25] Saro-Wiwa observed cynically that "African governments can ignore writers, taking comfort in the fact that only a few can read and write."[26] So, "the writer must be *l'homme engagé*," he wrote, "the intellectual man of action."[27] Launching MOSOP, establishing contact with Ogoni men and women in the towns and villages of their riverine territory, drawing on the emerging post-colonial strength of African literature in English, both oral and written, in his poetry, fiction, journalism and as the scriptwriter and producer of a popular television satire, Saro-Wiwa believed "the word is power," and that words were most powerful when they were "expressed in common currency."[28] Just as the Barbadian Kamau Brathwaite has said that all his poetry is about "*biggin-up* Black people & Africa,"[29] so Saro-Wiwa, in Chinua Achebe's words, represents "the highest standards of his country," the "true patriot" who accepted "nothing but the best for and from his people" without giving way "to superiority, despair or cynicism."[30]

"Literature," the Kenyan novelist Ngũgĩ wa Thiong'o has said, "should help us in returning to the notion of the sacred in life."[31] This is the literary stance Saro-Wiwa took concerning the exploration for oil in Ogoniland. "The land is a god and is worshipped," he wrote in *Genocide in Nigeria* (1992). Every aspect of the six kingdoms of Ogoniland[32] is considered "sacred" and "bound up intricately with the life of the community, of the entire Ogoni nation."[33] Animals and livestock "into which the human soul can be transmuted" are also considered sacrosanct, and Saro-Wiwa warned that "erring human action can desecrate them and therefore bring disaster upon the people who are regarded as their custodians."[34]

In Ogoniland, Mary Harvan has observed, "religion and everyday life revolve around the land which is at once the site of worship, production and reverence for the ancestors."[35] Saa, the owner of the earth, is the Ogoni deity who both gives and takes. The propagandist power of invoking a Saa-like 'curse' was not lost on Saro-Wiwa who "advanced the Ogonis claim to the land and its resources to deny ultimate power over them to the federal government and the oil companies."[36] At the launching of the One Naira Ogoni Survival Fund[37] in Bori on the 27th of February, 1993, Saro-Wiwa's "Call for Commitment" warned of "indigenous colonialism and environmental strangulation." He challenged the military regime in Abuja to revoke its "obnoxious, disgraceful and oppressive system" and warned not only of the death of the land but also of the "abject poverty, slavery, dehumanization and extinction"[38] of the people of Ogoniland. Saro-Wiwa concludes his beautiful poem "To Sarogua, Rain Maker," written during the Nigerian Civil War of 1967–1970, by desperately imploring the Ogoni god "Whose wink brings rain" to "Receive us, mighty one, With a cleansing shower."[39]

Issuing a caveat directed at Royal Dutch Shell and the Nigerian government, in an essay entitled "The Goose and the Prince," Saro-Wiwa encouraged Charles, the Prince of Wales, to "help save the peoples of the Nigerian Delta from the deprivation, disaster and doom which oil exploration has brought them in the last thirty years. Only in this way can the shame of the past be wiped out, the curse which the gods of the Delta have laid on the oil and the gas be lifted."[40]

Not only were the four ecological regions[41] of Ogoniland threatened, so also were the people by what was believed to be a Federal policy of genocide. "The Ogoni live in a nation," Saro-Wiwa argued, "which is determined to exterminate them and that Ogoni lives mean nothing whatsoever to Nigeria and other Nigerians," and, he added, "the murderous country would be pleased to see Ogoni territory rid of all its inhabitants so that the oil resources can fall freely into 'Nigerian' hands."[42]

Demanding the right to monetary compensation for lands 'legally'[43] seized and/or autonomy for Ogoniland, Saro-Wiwa wrote in *A Month and a Day*, "We repeated our demand for recognition of our political rights to self-determination, economic rights to our resources, cultural rights to the development of our languages and culture, and social rights to education, health, and adequate housing and to the representation as of right in national institutions."[44] This impassioned plea furthers Saro-Wiwa's hope that he had "started a movement which might transform Africa"[45] and corroborates Ngũgĩ's insistence that the "primary concern of literature" considers "the quality of human life" and "the right to live fruitfully, to live a life that increasingly makes human lives the meaning of life."[46] If the fictitious Dukana people in Saro-Wiwa's novel *Sozaboy* are like "beetles fighting slow slow with one big piece of smelling shit,"[47] then for Ogoniland and the rest of Africa "It is a good thing to fight" because "if somebody take your thing by force, if 'e want by force you to do something wey you no like do, then you fit fight am."[48]

In his influential polemic *Discourse on Colonialism,* the Martinican poet Aimé Césaire wrote that "no one colonizes innocently."[49] So when Royal Dutch Shell "entered into cultivated farms in the Bomu area [of Ogoniland] and mercilessly ploughed down acres and acres of cultivated crops; yams, maize, pepper, melons, fluted pumpkins, cassava, okra, garden eggs and other valuable crops," raising later accusations of 'ecological racism' as scientists began to realize the extent of the damage done to the already endangered ecosystems of the Niger Delta, MOSOP argued that similar extraction without compensation and cleanup, which incidentally began rather halfheartedly after Saro-Wiwa's death, would not be tolerated in the European oil fields or in the Gulf of Mexico. In Africa, however, the oil companies got away—literally—with murder, especially if a corrupt government turned a blind eye as it eyed a fine profit. When Royal Dutch Shell "destroyed by periodic out-flow of crude oil into [the] rivers and streams, which have killed off not only the mangrove trees, but fishes and crabs, mudskippers, oysters, shell-fishes, etc. on which the livelihood of the poorer people depends,"[50] the company's lack of an environmental policy went unchecked. Nigeria, under the dictatorships of Babangida and Abacha, in cahoots with Big Oil, simply held things back. Saro-Wiwa and MOSOP understood that the "blood money" extracted from "the very veins of [their] dying people,"[51] and that flowed into the barrel sized pockets of certain "lootocrats in Lagos,"[52] was underwritten by what Césaire called "a law of progressive dehumanization."[53] Colonial "Dismemberers,"[54] he argued, are licensed to commit a "gigantic rape of everything intimate, undamaged, [and] undefiled."[55] Saro-Wiwa, who "week after week . . .

made sure that the name Ogoni appeared before the eyes of [his] readers,"[56] accused Royal Dutch Shell of "complicity in genocide,"[57] and Abacha's government for causing a "state where peace or security does not exist," where "the life of a man is solitary, poor, nasty, brutish and short."[58]

"Dotted all over with oil wells, pipelines, flow stations, access roads and gas flares,"[59] Ogoniland, like the rest of independent and colonized Africa, was at the mercy, in Césaire's words, of "a machine for crushing, for grinding, for degrading peoples."[60] In "virtually every nation-state" in Africa, Saro-Wiwa wrote, "there are several Ogonis—despairing and disappearing people suffering the yoke of political marginalization, economic strangulation and environmental degradation, or a combination of these, unable to lift a finger to save themselves."[61]

While in detention in Port Harcourt Prison, Saro-Wiwa wrote his now famous song "Ogoni Star" (1995) to shake a fist in their name, to empower the will of his people in a few lines that include "dance the military guns to silence/dance their dumb laws to the dump/dance oppression and injustice to death"[62]

Ngũgĩ has argued that literature should "increase the spiritual wealth of a people."[63] In *The Trouble with Nigeria,* Chinua Achebe called on "all thoughtful Nigerians to rise up and reject those habits which cripple and inhibit" the country's "chances of becoming . . . modern and attractive."[64] Saro-Wiwa went further by addressing the kind of "thoughtlessness that has reduced Africa to beggardom, to famine, poverty and disease."[65] Many of the poems in *Songs in a Time of War,* like those of the Nigerian poet J. P. Clark-Bekederemo's *Casualties,* speak quietly the noise of war.

Whereas they recall the Ogoni dead during the secessionist conflict between the breakaway Republic of Biafra and the Nigerian Federation, the metaphor underpinning many of the twenty poems is the destruction of the land, its towns and its people. It is estimated that 30,000 Ogonis lost their lives in fighting for "a cause they barely understood."[66] The land is "filled with abomination."[67] In "Ogale-An Evacuated Town," Saro-Wiwa wrote, "Her legs apart like a cheap prostitute/Exposed, utterly exposed. Ogale is a ravished woman."[68] *Songs in a Time of War* serves as an Ur-text for Saro-Wiwa's literary war with Royal Dutch Shell and Nigeria. The resonant silences of Ogoniland's "plain of agony"[69] prepare us for the "infecting" of "God's good air"[70] by blow-outs and gas flaring. "The flames of Shell are hell/We bask beneath their light."[71]

Voicing concern in *This Darkling Plain* (1985) for "the harsh suppression of the minorites, [sic] the unabashed and remorseless exploitation of those handicapped by their numbers,"[72] Saro-Wiwa suffered the "shabby cruelty"[73] of Abacha's henchmen, one of whom Soyinka has denounced as "a gloating sadist"[74] who boasted he knew of 204 ways to kill a man.[75] Incident after incident was fabricated to discredit both MOSOP and Saro-Wiwa in particular. Finally, under orders from Aso Rock, four Ogoni moderates were murdered in an effort to turn MOSOP against itself. Reportedly aided by the bribing of witnesses by Royal Dutch Shell, Saro-Wiwa, together with Baribo Bera, Saturday Dbee, Nordo Eawo, Daniel Gbokoo, Barinem Kiobel, John Kpuinem, Paul Levura and Felix Nwate, were indicted under what was ambiguously termed a "criminal offence."[76] The Ogoni Nine were held in appalling conditions without trial and eventually sentenced to death by the same kind of kangaroo-court Saro-Wiwa had made fun of in his Swiftian novel *Prisoners of Jebs* (1988). The Ogoni Nine were executed in a botched hanging in Port Harcourt Prison on November 10th, 1995. Perversely, the bodies of the Nine were treated with lye and consequently denied a traditional Ogoni burial in the forest of their sacred land.

In *Sozaboy,* the words of Chief Birabee echo both Saro-Wiwa's hopes for Ogoniland and for Nigeria: "We shall all return home and we shall be in a new country where nobody will tief, there will be no hungry again, everything will be free, from water to food, to cloth to wear to medicine to lorry and licence."[77] This beautiful vision of Nigeria (and Africa) is dependent on winning the 'war.' Only then "will be life more abundant;" "If we do not win the war," the Chief adds, "what will happen."[78] In *Prisoners of Jebs,* Saro-Wiwa had described Nigeria as "a blessed country, endowed with a rich soil, mineral wealth, a strong and vigorous population . . . the main hope of the African continent, of Black people throughout the world . . . held out a dream which waited to be realized."[79]

Saro-Wiwa ends his poem "Dis Nigeria Sef" with the words "I don confuse well well at all/ Sake of dis I-love-I-no-love Nigeria./ I TIRE."[80] Exhausted but defiant, he nevertheless saw himself as "the conscience of society."[81] His war of words with "Big Motor and dirty House" continues in all corners of the Niger Delta, and in those places darkened by exploitation and greed all over the African continent. That Ken Saro-Wiwa gave his life for a cause that still speaks his words and bears his name is testament to the moral that ends the Ogoni folk tale, "The Power of Truth" in *The Singing Anthill* (1991): "Truth is powerful and is always victorious. It may lose sometimes, but only for awhile."[82]

NOTES

1. Kamau Brathwaite, *Ancestors* (New York: New Directions, 2001), 351.
2. Ken Saro-Wiwa, *The Singing Anthill: Ogoni Folk Tales* (Port Harcourt, Nigeria: Saros International, 1991), ix.
3. Brathwaite, *Ancestors*, 351.
4. Quoted in Wole Soyinka, *You Must Set Forth at Dawn: A Memoir* (New York: Random House, 2006), 427.
5. Ken Saro-Wiwa, "Dis Nigeria Sef," in *Songs in a Time of War* (Port Harcourt, Nigeria: Saros International, 1985), 40.
6. Also involved were Agip, Chevron, Elf and the Nigerian National Oil Corporation.
7. Wole Soyinka, *The Open Sore of a Continent: A Personal Narrative of the Nigerian Crisis* (New York: Oxford University Press, 1996), 27.

8. Ibid., 133.
9. Karl Maier, *This House has Fallen: Nigeria in Crisis* (Boulder, Colo.: Westview, 2002), 78.
10. Ken Saro-Wiwa, *A Month and a Day & Letters* (Banbury, England: Ayebia, 1995), 50.
11. Ibid., 101.
12. Soyinka, op cit, 151. The historian Toyin Falola notes wryly that Major General Sani Abacha's "era was marked by despotism and economic decline" whereas Major General Ibrahim Babangida had "a desire both to be a kingmaker and later a king." *The History of Nigeria* (Westport, Conn.: Greenwood Press, 1999), 231, 235.
13. Ken Saro-Wiwa, *Genocide in Nigeria: The Ogoni Tragedy* (Port Harcourt, Nigeria: Saros International, 1992), 56.
14. William Wordsworth, *The Prelude,* Book III, 82.
15. Saro-Wiwa, *Genocide in Nigeria,* 59.
16. William Wordsworth, *The Prelude,* Book II, 82–85, 104.
17. Quoted in Ike Okonta and Oronto Douglas, *Where Vultures Feast: Shell, Human Rights, and Oil* (London: Verso, 2003), 172.
18. Primitive rebel is a Robin Hood-like hero who spoke and acted for and against a group of unrepresented people against those who would disrupt the "social balance of the kinship society by turning some kinds into 'rich families and others into 'poor', or by disrupting the kin itself." See Eric Hobsbawm, *Primitive Rebels: Studies in Archaic Forms of Social Movement in the 19th and 20th Centuries* (New York: Norton, 1965), 4.
19. Percy Shelley, *Shelley's Poetry and Prose*, ed. Donald H. Reiman and Sharon B. Powers (New York: Norton, 1977), 310, 372.
20. *Genocide in Nigeria,* 56.
21. Op. cit., 255.
22. Ken Saro-Wiwa, *A Month and a Day & Letters* (Banbury, England: Ayebia, 1995), 56.
23. Quoted by Frank Schultze-Engler in "Civil Critiques: The Satire and Politics of Democratic Transition in Ken Saro-Wiwa's Novels" in *Ogoni's Agonies: Ken Saro–Wiwa and the Crisis in Nigeria,* ed. Abdul Rasheed Na'Allah (Trenton, N.J.: Africa World Press, 1998), 291.
24. *A Month and a Day & Letters,* 126.
25. Ken Saro-Wiwa, *The News* (May 17, 1993). Quoted in *Ogoni's Agonies,* 109.
26. *A Month and a Day & Letters,* 55.
27. Ibid.
28. Ibid.
29. "Kamau Brathwaite Talks With Emily Allen Williams About Four Decades of Critical Response," in *The Critical Response to Kamau Brathwaite* (Westport, Conn.: Praeger, 2004), 306.
30. Chinua Achebe, *The Trouble with Nigeria* (London: Heinemann, 1983), 15–16.
31. In an interview given to Lee Nichols (1996) in Reinhard Sander and Bernth Lindfors, eds., *Ngũgĩ wa Thiongo'o Speaks: Interviews with the Kenyan Writer* (Trenton, N.J.: Africa World Press, 2006), 381.
32. They are Babbe, Eime, Gokana, Ken-Khana, Nyo-Khana and Tai.
33. Ibid., 12.
34. Ibid., 12–13.
35. Mary Harvan, "Religion, Ken Saro-Wiwa and the Ogoni Movement" in *Mapping the Sacred: Religion, Geography and Postcolonial Literatures,* ed. Jamie S. Scott and Paul Simpson-Housley (Amsterdam: Rodopi, 2001), 244.
36. Ibid., 243.

37. "All Ogoni men, women, and children, including newborn babies" Saro-Wiwa wrote, "will contribute to the Fund as a statement of their will to survive as individuals and as one indivisible nation." See *A Month and a Day & Letters,* 99.
38. *A Month and a Day & Letters,* 97.
39. Ken Saro-Wiwa, *Songs in a Time of War* (Port Harcourt, Nigeria: Saros International, 1985), 30. Instead of a rejuvenating downpour, however, acid rains still falls on Ogoniland. Not only is Big Oil and its exploitation of one of the world's fragile ecosystems 'cursed' as it deals to this day with oil bunkering and random acts of violence and kidnapping of its employees, so also are the approximately half a million Ogonis, some of whom still believe that the finding of oil on their "God-given land" was itself "the greatest curse that ever happened"; *Genocide in Nigeria,* 77. In the short story "Night Ride," Saro-Wiwa's narrator says of himself, "He cursed the earth for spouting oil—black gold, they called it. And he cursed the gods for not drying up the oil wells." See Ken Saro-Wiwa, *A Forest of Flowers* (Harlow, England: Longman, 1995), 112–113.
40. Ken Saro-Wiwa, *Similia: Essays on Anomic Nigeria* (London, Lagos & Port Harcourt: Saros International, 1992), 74–75.
41. They are the coastal sand barrier islands, the ecozone of West African lowland equatorial monsoon, the West African freshwater alluvial equatorial monsoon and the West African brackish-water alluvial equatorial monsoon. See *Where Vultures Feast,* 61–62.
42. Op. cit., 43.
43. In "Ethnic Minorities and the Nigerian States: The Ogoni Struggle after Ken Saro-Wiwa", in *Before I am Hanged: Ken Saro-Wiwa: Literature, Politics and Dissent,* ed. Onookome Okome (Trenton, N.J.: Africa World Press, 2000), Felix Atpan writes, "Under Nigerian Petroleum Laws, the federal government appropriates the right over oil and gas and gets all revenue accruing from its production. Thus, the primary responsibility of developing the oil-producing area is that of the Federal government. Under the same law, the oil companies do not have a legal responsibility to develop the oil-bearing areas. Their statutory obligation is to pay royalties and taxes to the federal government. This apparently explains Shell's indifference to the plight of the Ogoni people."
44. *A Month and a Day & Letters,* 62.
45. Ibid., 92.
46. Op. cit., 281
47. *Sozaboy,* 157.
48. Ibid., 17.
49. Aimé Césaire, *Discourse on Colonialism*, trans. Joan Pinkham (New York: Monthly Review Press, 1972), 17.
50. *Genocide in Nigeria,* 46.
51. Ibid., 48
52. Ibid., 85.
53. *Discourse on Colonialism,* 49.
54. Ibid., 53.
55. Ibid., 60.
56. *A Month and A Day & Letters,* 65.
57. *Genocide in Nigeria,* 57.
58. *Genocide in Nigeria,* 59.
59. Ibid., 76.
60. *Discourse on Colonialism,* 60.
61. *A Month and a Day & Letters,* 123.

62. Quoted in *Where Vultures Feast,* 183.
63. Op. cit.
64. Ibid., 2.
65. Ken Saro-Wiwa, *The Prisoners of Jebs* (Port Harcourt, Nigeria: Saros International, 1988), 58.
66. "Thoughts in Time of War," in *Songs in a Time of War,* 24.
67. "Corpses Have Grown," in *Songs in a Time of War,* 18.
68. "Ogale-An Evacuated Town," in *Songs in a Time of War,* 29.
69. "Epitaph for Biafra," in *Songs in a Time of War,* 33.
70. Ibid., 34.
71. Quoted in *Genocide in Nigeria,* 71.
72. Ken Saro-Wiwa, *On A Darkling Plain: An Account of the Nigerian Civil War* (Port Harcourt, Nigeria: Saros International, 1989), 11.
73. *The Open Sore of a Continent,* 149.
74. Ibid., 4.
75. See *This House has Fallen,* 106.
76. Quoted in *Feasting with Vultures,* 157.
77. *Sozaboy,* 154–155.
78. Ibid., 155.
79. *Prisoners of Jebs,* 146.
80. *Songs in a Time of War,* 44.
81. From Saro-Wiwa's preconviction statement, quoted in *Where Vultures Feast,* 207.
82. *The Singing Anthill: Ogoni Folk Tales,* 112.

9 Religion in the Development of Environmental Ethics in Nigeria
Niger Delta in Perspective

S. O. Ijaola and T. T. Asojo

INTRODUCTION

The controversial root of current environmental problems has generated heated arguments among various scholars with their critical views channeled through political means. Mackinnon noted that,

> The fact that we differ in our evaluation (of environmental problems) is not a problem in itself for the anthropocentric position or for attempts to use cost-benefit analyses to aid our decision making. We argue, and changes [sic] our minds and sometimes we agree to disagree, then, we find political ways to negotiate these differences.[1]

The issue of environment, especially its influence on the economy, has made it an imperative concern for the political authorities of most nations. The scourge of environmental degradation particularly in developing countries was highlighted as one of the main Millennium Developmental Goals (MDG) in the September 2000 Summit of the United Nations where world leaders gathered to set a time and measurable targets to combat the ills of underdeveloped nations including poverty, hunger, disease, illiteracy and environmental degradation.[2]

Environmental deterioration has been observed as a major cause of poverty in Nigeria, and the Niger Delta area is a concise instance where environmental exploitation and degradation contributes to violence and entrenched poverty. The International Fund for Agricultural Development (IFAD) 2007 states that, "Poverty and violence are often closely interconnected. . . . In the Niger Delta, where the oil industry is based, a vigorous trade in stolen oil has led to a serious break-down of law and order in that area."[3] It is obvious that environmental problems reinforce food insecurity, which automatically paves the way for prevailing poverty. Nigeria is presently witnessing environmental crises such as global warming; ozone depletion; pollution of the land, water and air; desertification; and floods with unparalleled loss. Nevertheless, the environmental deterioration can be remedied, only if there is a unified approach and more generally by a painstaking effort to burrow into the relationship of nature and mankind; a project that must involve all necessary stakeholders among which religion plays a unique role. To this end, the chapter advances the argument that religion with the aid of contextual theology is a veritable tool for the realization of the development of proactive and efficient environmental ethics in Nigeria.

PERSPECTIVES ON ENVIRONMENTAL ETHICS

Several perspectives on the issues of environment and ecology have emerged over time revealing diverse human attitudes to the environment. This section advances the argument that religion influenced the differing perspectives of most people as regarding the human relationship with the environment. Hence the blame on religion for being the cause of the environmental problems we are experiencing in the world today. The Newtonian mechanistic worldview is also indicated to be the bane of ecological problems, yet religion cannot be completely excused because it prepared the ground from the mechanistic worldview of modern science. It is common for industrialization to be labeled as the cause for environmental degradation; however, there are specific Western religious ideologies that also contributed to environmental irresponsibility.

The term environmental ethics has been used recently to indicate a wide range of issues that involve the relationships of human kind to the animate and inanimate world. There are revived perspectives on environmental ethics. From a theoretical point of view, Galleni, in his article "Theology of Nature as a Contextual Theology: Ethical implication," described human behaviors through hosting in relation to nature by the following:[4] the anthropocentric perspective, from which strong anthropocentric view (nature's value judged by human its use for man), and the weak anthropocentric, which includes both the preservationist and the conservationist (nature's value judged by human but with consideration for preservation); the non-anthropocentric perspective (nature has its intrinsic value to be considered and respected by humans); and the deep-ecology perspective (nature's value to be given equal stance as human's value).

Among these three perspectives, the anthropocentrism and non-anthropocentrism are the two opposing ends. Anthropocentrism is the human definition of the role of nature in the world. It is a human's judgment of nature's value, whereas non-anthropocentrism describes the definition of nature's role and value in the world independent of human's definition. Strong anthropocentrism affirms the absolute supremacy of man over nature. This is described by Vicente Beller Capella in "The Relationship of Human Kinds and the Earth in the Light of the Pentateuch."[5] It gives credence to technocratic ecophilosophies that project human behavior as the only criterion for taking decisions concerning the world.

It follows therefore that the optimism of modernity with respect to the ability and right of humans to tend the world and utilize its resources depends on them. As a consequence, the value of nature corresponds to the market utility that can be gained from environment's exploitation. This perspective, however, has lost influence as more information concerning the exhaustion of nonrenewable resources and as developmental models have elaborated the consequences of man's intervention in the environment. The consequence of which is the impossibility of unlimited development of such markets.

The preservation perspective is linked to the necessity for nature's preservation. In this viewpoint, the destructive attitude of man toward nature is mitigated by the need to preserve some portions of nature (such as parks, natural reservation and protected areas) that have importance exceeding strictly economic value. In Lodovico Galleni's opinion, this importance can be a source of physical recreation (nature gymnasium),[6] of spiritual concentration (nature Cathedral) or a school of solitude and a challenge to human arrogance, i.e., a source of psychological transitions. In this way nature is seen as encouraging the development of moral character by stimulating higher thoughts and higher levels of humanity. In cases where nature's natural beauty is preserved for aesthetic pleasures as found in museums or art galleries, the primary aim is to enhance human's value of nature.

The conservation perspective is engendered toward a rational use of nature owing to its economic use for future human generations and for protection. This requires a deep knowledge of the environment, which can be acquired by ecologists, etiologists and biological and cultural anthropologists. In this view man is still the lord and controller of nature. Yet the preservation and conservation perspectives are therefore considered to make a crucial point for weak anthropocentrism. Vicente Beller Capella described this view as humanistic ecophilosophies, which consider that the recognition of the value of humanity does not imply that the rest of nature has an instrumental role. Here there is a concern for sustainable development that takes into account that natural resources are for humankind but are also limited. So humankind has to look after its reposition and conservation.

The Non-anthropocentric perspective is described as anti-anthropocentric or 'for-the-respect of life.' The view affirms that nature has a real value not lower than humankind's and it must be preserved independently of its satisfaction of human needs. The positions of the perspective include the concept of suffering; that is, if an entity can feel pain and suffering it deserves moral consideration. The presence of such ability in a non-human being should give it the rights similar to human egoism.[7] According to this idea morality applies to all living entities including those unable to feel pain or pleasure.[8] These positions rest on the three principles as identified by Galleni, first that life is a continuum and we cannot distinguish between higher and lower forms of biochemical processors and behaviors.[9] Second, they have their peculiar interest to pursue, and man must respect their integrity even if this goes against his own interests.[10] Third, all natural entities, both living and inanimate, bear real value that makes them holder of rights such as not to be altered, degraded or used.[11] Man, therefore, has the obligation to respect these rights and must protect the environment; only by this way will he achieve his good.

The deep-ecology perspective is the view that supports a biocentrism or egalitarian biologism. In other words, all species have the same right to develop according to their nature. Deep ecology affirms that men should have interior bonds with non-human beings and must improve these relations through practical experiences.[12] The main problem is that it is very difficult to find a way to propose unification or still, a common ground of discussion among the widely differing perspectives.

The relationship of theology to the modern ecological crisis became an intense issue of debate in 1967, following the publication of the article, "The Historical Roots of our Ecological Crisis" by Lynn White. After detailing aspects of the ecological crisis, White proceeds to fault Christianity for propagating an anti-ecological understanding of the relationship between humanity and the environment. His criticism is twofold. "Christianity is the most anthropocentric religion the world has seen . . . [It] not only established a dualism of man and nature but also insisted that it is God's will that man should exploit nature for his proper use."[13] White acceded that his criticism applies to Western Christianity in the post-scientific revolution era. In this regard, it may be more accurate to indicate that White is not necessarily criticizing Christian theology per se but Christians who improperly imbibed the Cartesian/Newtonian worldview. "The success of Newton's mechanistic worldview led to a significant religious development"[14] . . . "whilst he scrupulously tried to avoid allowing religious prejudices to colour his scientific theorems," and considered that 'in the last analysis, the realm of science was dependent on the God of religion and led the reverent mind to a fuller assurance of his reality and a readier obedience to his commands"[15] This happenstance between Newton science and religious conceptions therefore corroborated the view that

> The association between Christianity and the rise of modern science is often regarded as being more than historical coincidence as Christian belief in the rational ordering of the universe, and the predictability of nature's laws, encouraged the investigation of the regularities underlying causative and motive structure of the natural order.[16]

It was from this context that the industrial revolution, which gave room for the Victorian era and the use and abuse of natural resources, began. "Cruelty towards animals, both for scientific research and for entertainment, was culturally acceptable, and the Victorian era, zeal for hunting and collecting wild animals resulted in the extinction of many species."[17] This background gives credence to White's conclusion in his article with a challenge to theologians to present an alternative Christian view that is ecological, "since the roots of our trouble are so largely religious, the remedy must also be essentially religious."[18]

Paul H. Santmire's *The Travail of Nature: The Ambiguous Ecological Promise of Christian Theology,* argued that "Christian thought is both promising and not promising for those who are seeking to find solid traditional foundations for a new theology of nature."[19] In his survey of the Christian tradition and its diverse attitude towards nature, Santmire expresses the thought of a number of key Christian theologians including Iranaeus, Origen, Augustine, Aquinas, Bonaventure, Dante, St. Francis of Assisi, Luther, Calvin and Barth.[20] As a result of his analysis, Santmire identifies two opposing theological motifs that are interwoven like brightly colored thread through the historical tapestry of Christian thought. He describes these two motifs as the spiritual motif that expresses a fundamental theological bias in the characteristics of God as a being separate from or transcendent to the world but who chooses to intervene in its affairs at will.[21]

Santmire quotes a well-known phrase from Augustine's soliloquies as a basic expression of this motif. "I desire to have knowledge of God and the soul of nothing else whatsoever."[22] Ultimately, nature is affirmed as 'good' only in its ability to embody spirits, which is the final measure and end of all theological enquiries for the spiritual motif.[23] Santmire therefore took a motif for his ecological theology position that is in contrast to the spiritual motif, as he submitted that

> The human spirits witnessed in the world of nature and on the desire of self consciously embodied selves to celebrate God's presence in, with, and under the whole bio-physical order. '[E]cological' in ecological motif is meant to express the systematic interrelationships between God, humanity, and the natural world.[24]

This theological thinking arose from the roof metaphors, which are the metaphor of ascent, the metaphor of fecundity and the metaphor of 'migration to a good land.'[25] However, inherent in this ecological metaphor, Santmire does point out that what is crucial is the ethical relationship to the land that one chooses to embrace, for as we know, there are many. Thomas Berry in "The Mythical-Cosmological Dimension of Environmental Ethics," claims that

> We need not a human answer to earth function in its own ways; we need only to listen to what the earth is telling us and for this hearing to occur, humanity needs a better story by which to live: we need a story that will educate us, a story that will heal, guide, and discipline us.[26]

In the story that Berry proposes, we need to imagine ourselves less as "a bridge on the earth or in the universe than a dimension of the earth and indeed of the universe itself."[27] He also concluded that the estrangement from nature grows of two flawed myths: the religious myth of a transcendental redemption earth and the secular myth of ultimate fulfillment on earth through increased technology and rationality, which he named as the 'industrial myth.'[28] The resolution then will naturally stem from the ascendance of a new myth. Thus, Berry agrees with the Lynn White thesis that the answers to ecological problems arise from particular religious worldviews, which are constructed from particular myths. Berry therefore proposes a 'mystique of the land' to counteract the industrial mystique.

Jürgen Moltmann in *God in Creation* developed a thrust for an ecological theology. One of his guiding ideas for an ecological doctrine is God's immanence in the world. He indicates that new ecological thinking about God must no longer center on the distinction between God and the world but on the recognition of the presence of God in the world and the presence of the world in God.[29] This theology is distinct from the pantheism of contemporary ecological thought. However, Osborn noted that there is a tendency among Christian Greens (ecological-oriented believers)[30] to overemphasize divine immanence at the expense of His transcendence.[31] He commented further that what is necessary is a doctrine of divine transcendence so radical that it will necessitate belief in the divine immanence.[32] Moltmann refers to this as paratheism. Nonetheless, the difference opined by ecologists and theologians boils down to the necessity of univocal environmental ethics, which is largely accomplishable through the instrumentality of religion.

It therefore appears clear religion is not excluded from the environmental debate that is becoming increasingly interesting among various professionals and scholars. Religion, especially Western Christianity, has been singly accused of propagating the ongoing exploitation of nature, although further arguments rather differentiate the biblical interpretation from the Newtonian mechanistic orientation that pervaded the Western world in the rise of modern science, which was responsible for the colorized interpretation of Gen. 1:26 and 28. Yet because Christianity is the way of life of modern Western society and the ground on which modern science develops, consequently, its contribution to environmental and ecological issues remain inevitable.

THE VALUE OF CONTEXTUAL THEOLOGY AND THE NIGER DELTA CONTEXT

Contextual theology is a potent tool for achieving a tremendous feat in the development of environmental ethics. The main discovery of twentieth-century science is the awareness that we are living on a small planet of limited resources.[33] Through the theological genius Moltmann it is argued that "The final benediction of Genesis 1: 25–31 is a fundamentally revised possession form to Leviticus 25–26 where God's final possession of earth is strongly affirmed. This a clear example of exegesis related to contextual theology."[34] Therefore, this section advances the idea that it is very important for Christian theologians in the African environment to explore an exegetical understanding of the scripture within the context of Africa, especially in the Niger Delta region of Nigeria, to develop biblical environmental ethics that will assist Christians and Churches within that region.

The contemporary context of environmental ethics presents a situation of exploitation and conservation of nature. On the other hand, the scientific context is not only that of biological evolution but also of a global approach referred to as the biosphere-centric theory.[35] The problem here is that the context of nature is mainly related to the conservation of nature; whereas the main problem regarding ethics is when we pass from a reductionist approach to a holistic approach. We have not only a quantitative change but also a qualitative one. It will then be necessary to look carefully at what kind of new problems are posed mainly by the global stability of biosphere.[36]

Environmental ethics is therefore not solely the ethics of the conservation of endangered species anymore, not of the single ecosystems, not of peculiar areas, but it is the ethics of biosphere stability. In the global perspective, environmental ethics is no longer an ethics regarding individual behavior, but the behavior of all the noosphere (thinking beings) in respect to the whole biosphere.[37] It is necessary for the biosphere to maintain equilibrium, and this is necessary for the noosphere too. The noosphere can act in a rational way to maintain the equilibrium and to avoid the risk of an alteration of the equilibrium owing to human or natural causes.[38] The emergence of the thinking sphere is an advantage also for the biosphere, because its tasks are accomplished in cooperation with the rational acting noosphere. Weak anthropocentrism is possible only with a global approach, but within this approach is injected the concept of symbiosis.[39]

The survival of biosphere is warranted by its task when it is considered as a whole and the survival of the noosphere in symbiotic relationship with the biosphere. Every artificial change in the structure of living beings, such as those proposed by biotechnologies, has to answer the basic question of their relatedness to general stability. This is a new path in environmental ethics opened by a biosphere-centric approach.[40] In this perspective all the discussion about biotechnologies have to give an answer to questions relating to the stability or instability of biosphere. And do we have the techniques to gather information about the actual effect of biotechnology on global environmental stability?

The Nigerian context presents an example of the unchecked destruction of the biosphere, leading to various problems such as restiveness of the communities that made up the Niger Delta region. Desertification, destruction of the aquatics in the coastal region, flood, health hazards and depletion of the ozone are obvious results of environmental exploitation in the area:

> It is a truism that there has been a progressive loss or reduction in the quality of our physical environment as well as human health since the industrial revolution in the 19th century. In other words, the quality of human health and life expectancy has drastically reduced in recent [years and] negatively altered the quality of the environment.[41]

Taking the Ogoni land as an instance of a context in Nigeria where Shell has been drilling oil since 1958, the people of Ogoni land who had pipelines built across their farmland and in front of their homes, suffered endemic oil leakages from these pipelines and were forced to live with constant gas flaring. This environmental assault smothered the land with oil, killed masses of fish and other aquatic life and introduced devastating acidic rain to the land. When crude oil touches the leaves of a yam or cassava, or any kind of economic tree, it dries immediately. "Oil exploration and exploitation have over the last four decades impacted disastrously on the socio-physical environment of Niger-Delta oil bearing communities."[42] It is added that 70.7% of the 779 people interviewed on the socioeconomic impact of oil in Nigeria identified farmland pollution as a major problem[43] in the Ogoni land.

The underlying cause of the environmental destruction is what is tagged as a 'Business as usual'[44] attitude because the main objective of every business venture is profit maximization. The strong anthropocentric perspective is obviously dominant in the Niger Delta area with the presence of the more dominant transnational oil companies, Shell, Mobil and Chevron. However, the environmental problem has generated many controversies. For instance

> Environmental claims are increasingly incorporated into the rhetoric of insurgency and need to be independently addressed. Locals have long complained that spilled oil from deteriorating decades-old pipelines has devastated fishing, although over fishing is also to blame. Oil companies insist that the vast majority of spills that have occurred in recent years are the result of sabotage by oil thieves and other groups trying to extort compensation payments.[45]

The crisis situation in the Niger delta centers on the approach to the environment. Nigeria as a nation presently relies solely on oil for its national income. This of course should stimulate the country to develop a strong environmental policy that will encourage environmental ethics in the society to protect the Ogoni environment from exploitation and antecedent problems such as oil spillage on the farm land, bunkering and violence. Several youths have resorted to militancy, kidnapping and oil theft for survival because of the poor compensation for land exploration. It is therefore imperative for environmental ethics to be univocally implemented in Nigeria, and religion must play a role in developing environmental ethics along with several other organizations in Nigeria.

Religion and environmental ethics in the Nigerian context becomes a necessary focus in this study owing to emerging environmental problems (basic problem of the environment, e.g., climatic problem, drought and others) and ecological problems (resulting from human relationship with the environment, e.g., disease, oil spillage). But it seems that religion is yet to play a vital role in the development of environmental ethics. However, it is commonplace that African culture promoted by African religion gives a considerable respect to nature both in their worship and practices, which is evident in groves and sacred forests. African culture made certain taboos to regulate killing of fish and animals indiscriminately, felling young trees and others. It is therefore obvious that African religions have a good sense of eco-theology.

Western Christianity and theology is responsible for unguided eco-exploitation in the world as opined by Lynn White. However, with the challenge thrown by ecologists, biologists and environmentalists to theologians on the need to develop an environmental friendly theology that has concern for a balanced relationship between humans and non-humans should propel religious adherents in Nigeria to rise to the task of developing environmental ethics. It is necessary for African Christianity to emerge with a strong eco-theology just as in orthodox traditions of Eastern Christianity. Saint Francis of Assisi and other orthodox traditions were environmental friendly in their expressions of Christian faith as opposed to Protestantism and Catholicism that permitted blatantly disregard for the environment.[46] Therefore, Lynn White as observed by Jorgen Moltmann is partially correct for blaming Western Christianity for the world's ecological exploitation.[47]

The Jewish perspectives presented by Preston Hunter observed that the "modern environmental movement is a growing social force, akin in many ways to traditional world religions."[48] He added that "the call for an abandonment of traditional beliefs and practices in order to effectively respond to the current environmental crisis is unnecessary for contemporary Jews; because Judaism as a traditional belief and practice do not conflict with modern, scientifically sound environmental practices."[49] Traditional Judaism espoused philosophies and practices that coincide remarkably well with those prescribed by the modern environmental viewpoint.[50] He also accepted that Jews and Christians today are to bring human hubris under some sort of control based on the sacredness of nature and taking into account the way God's creation actually works.[51] He also posited that central to the Jewish treatment of deforestation and other ecological ethics is the principle called *ba tashit*,[52] meaning "thou shall not destroy."

African theologians need, therefore, to respond to the need for an environmental ethics crusade understanding the negative impacts its exploitation has on its adherents. In this light, Nigerian religious groups must come up with an agenda to play a crucial role in the development of environmental ethics along with their advocacy for justice and human rights in the Niger Delta.

> The churches worldwide have continued to work for justice. . . . In the post cold war period, there has been a marked increase in human rights violations, particularly on the African continent. Underlying these human rights violations are concerns about the deepening poverty of the mass of African people and the marginalization of the continent as a whole in light of globalization and economic restructuring. Unit III of the World Council of Churches, over a period of time, has monitored the events and developments in Nigeria in general and in Ogoniland in particular. It has kept itself informed of the situation through its member churches and partners in the country.

The purpose and objectives of the meeting, besides sharing and exchange of information, was, among others, sharing with the Nigerian churches the effect of socioeconomic and political realities on Nigerians' constitutional rights; helping local assemblies through capacity building to engage them in upholding justice for the masses in the country; providing an avenue for discourse among participating churches in the vanguard of defending justice; and creating a free flow information channel among member churches.[53]

These laudable interventions of religious institutions without a conscious and long-term program toward campaigning for and the development of biblical environmental ethics within the Nigerian context may only have a temporal effect on the people because people are not educated about the biblical perspective of the relationship that exists between them and their environment. The fact that environmental consciousness is now a global concern should stimulate a true local participation from local churches and this should begin by contextualizing biblical ethics in terms of man's responsibility to his habitat. Some other interventions could champion the course of developing environmental ethics in Nigeria through workshops and seminars that aim at enlightening the public on environmental problems and proffering contextualized biblical solutions. The local assemblies and denominations have the opportunity of promoting environmental ethics through institutions like the Christian Association of Nigeria (CAN) as well as advise, collaborate and partner with government, foreign organizations, private bodies and nongovernmental organizations (NGOs) with environmental sustainability as the common interest. It must also be emphasized that Christians and churches that practice environmental care be seen as important acts of worship and service to the Lord.

It is evident that religion has a crucial role to play in the development of environmental ethics in Nigeria. The Niger Delta crisis clearly shows how much destruction of lives an environmental crisis can cause for both the biosphere and the noosphere. It is therefore very pertinent that the government of Nigeria cease slacking in arresting the situation. However, the task of encouraging environmental responsibility must be carried out with the understanding of current global ecological trends on the issues of environment and ecology. The influence of technology and science on the ecosystem cannot be ruled out in present society. Just as with the rise of modern science, this may likely be worse for developing countries that do not have the facilities and arrangement for recycling some of technological by-products. It is prudent for the government to consider critically the negative impact of technological products before allowing use in the country. The question should be can Nigeria properly manage the waste of these products?

The challenges posed by ecologists, biologists and environmentalists to religion need to be at vanguard in the understanding of the context for the development of environmental ethics. Policies are not enough. There is a need for ethics. It is the people who are the central participants in the process. Government efforts to formulate policies must, however, be met with good environmental ethics, which cannot be done without the tool of contextualizing biblical views on environment and ecology. It is imperative therefore that religious institutions should endeavor to cursorily study and strategize their theological skills, stewardship of influence and their institutional roles to achieve the development of environmental ethics in Nigeria.

NOTES

1. Barbara Mackinnon, *Ethics: Theory and Contemporary Issues* (New York: Wadsworth Publishing Company, 1998), 392–393.
2. African Development Bank (ADB) Annual Report, 2007.
3. International Fund for Agricultural Development (IFAD) Report, 2007, http://www.rural/povertyportal.org/english/religions/africa/nga/index.htm. Accessed September 24, 2008.
4. Lodovico Galleni, "Theology of Nature as a Contextual Theology: Ethical Implication," *Studies in Science and Theology* 9 (2004): 31.
5. Puig liorenc, "The Relationship of Human Kind and the Earth in the Light of Pentateuch," *Studies in Science and Theology* 9 (2004): 29.
6. Galleni, "Theology of Nature as a Contextual Theology," 41.
7. Liorenc, 53.
8. Galleni, "Theology of Nature as a Contextual Theology," 34.
9. Ibid., 35.
10. Ibid., 35.
11. Ibid., 37.
12. Ibid.
13. Ibid.
14. Alister E. McGrath, *Science and Religion: An Introduction* (Boston, Mass.: Blackwell Publishers, 1999), 18.
15. Hillary Marlow, *Biblical Prophets and Contemporary Environmental Ethics* (Oxford : University Press, 2009), 53–56.
16. Michael S. Northcott, *The Environment and Christian Ethics* (Cambridge: Cambridge University Press, 1996), 57.
17. Marlow, *Biblical Prophets and Contemporary Environmental Ethics,* 54–55.

18. Galleni, 39.
19. Paul H. Santmire, *The Travail of Nature: The Ambiguous Ecological Promise of Christian Theology* (Minneapolis, Minn.: Fortress Press, 1995). http://www.counterbalance.org/enviro/hpaul-frame.html. Accessed September 30, 2008.
20. Ibid.
21. Ibid.
22. Ibid.
23. Ibid.
24. Marlow, *Biblical Prophets and Contemporary Environmental Ethics,* 53–56.
25. Santmire, *The Travail of Nature.*
26. Ibid.
27. Thomas Berry, "The Mythical-Cosmological Dimension of Environmental Ethics." http//www.schorlas.edu/Thomas/mythical.html. Accessed December 2, 2008.
28. Ibid.
29. Jurgen Moltmann, *God in Creation* (Minneapolis, Minn.: Fortress Press, 1993).
30. Osborn (1990), http.//www.p.s.a.net. Accessed December 2, 2008.
31. Ibid.
32. Moltmann, *God in Creation.*
33. Galleni, "Theology of Nature as a Contextual Theology."
34. Ibid.
35. Ibid.
36. Galleni, "Theology of Nature as a Contextual Theology," 37.
37. Ibid.
38. Galleni, "Theology of Nature as a Contextual Theology."
39. Ibid.
40. Ibid.
41. Prince C. Mmon, "Environmental Management System: Constraints To and Strategies for Implementation and Compliance by Corporation in the Developing Countries," (Nigeria, 2006). www.ui.edu,ng/researchgeo.htm. Accessed December 10, 2008.
42. Owabukeruyele Mnon, "Hydrocarbon Exploitation, Environmental Degradation and Poverty in the Niger Delta Region of Nigeria" (2000), 4 Sweden Lund University, LUMESProgram. www.wado.org./environment/vol companies/sehllcommunities/claudeake. Accessed December 12, 2008.
43. Mnon, "Environmental Management System," 4.
44. Ibid.
45. "Fuelling the Niger Delta Crisis," Dakar, Senegal and Brussels, Belgium (2006). www.crisisgroup.org/en/africa/west-africa/nigeria/. Accessed December 12, 2008.
46. Galleni, "Theology of Nature as a Contextual Theology," 39.
47. Moltmann, *God in Creation.*
48. Hunter Preston, "Jewish Perspective." www.jewish/perspective. Accessed December 13, 2008.
49. Ibid.
50. Ibid.
51. Ibid.
52. Ibid.
53. World Council OF Churches, "Report of the international Strategy Meeting on Nigeria," November 20–24, 1996, http://www.africaaction.org/. Accessed November 24, 2008.

Part III

Environmental, Sociocultural and Political Dimensions

10 From Oil Rivers to the Niger Delta

The Paradoxes of Domination and Resistance in Colonial Nigeria

Charles Ukeje

[Few] aspects of African history have generated as much interest or undergone as dramatic a reinterpretation as the study of resistance to colonial rule.[1]

The impulse for this paper derives from an attempt to revisit critical aspects of the historiography of the Niger Delta that might serve as a template for understanding the complex realities in the region today. My concern, for the most part, goes beyond taking up the challenge of "reinterpretation" for its own sake but originates from an intellectually uneasy concern that a corpus of pioneering works on the historiography of peoples and societies in Nigeria's delta have yet to be sufficiently deployed to tackle the perennial disorders that have enveloped the region, especially since the 1990s. Despite the strategic importance of the delta region for Nigeria, and also the context of global geopolitics of hydrocarbon oil, further critical illuminations on the complex genealogies of the present are due. This concern is heightened by the fact that current scholarship seems less aware of, engaging with and profiting from the seminal works on the region by prominent first-generation historians of Nigerian extraction, notably Kenneth Dike, Tekena Tamuno, Ebiegberi Alagoa, J. C. Anene and Obaro Ikime, to mention a few.[2] My aim is to capture what key insights those pioneer scholars offer in understanding the prolonged and enriching traditions of domination and local resistance in the Niger Delta during the colonial era and thereafter.

Before it was rechristened the Niger Coast Protectorate, the Oil Rivers had the earliest and longest contact with Europeans spanning over six centuries, making it central to the making of the modern entity called Nigeria. As the first port of entry where European incursion into the hinterlands was anchored, it encountered much earlier than other parts of the west coast of Africa a succession of European adventurers, missionaries, merchants and mandarins. During much of that period, the coastal delta kings and their peoples successfully served as middlemen during the different phases of the trans-Atlantic trade in slaves and later palm oil. This *longue durée* offers enriching perspectives on the changing political economy of the delta and how this influenced the pattern and trajectory of domination and resistance, which punctuated the colonial (and post-colonial) periods.

This chapter solicits a more nuanced engagement with and understanding of history, politics and society in the colonial delta, insisting that much of the enduring puzzles and contradictions of domination and resistance in the contemporary delta might come into sharper relief by closely interrogating history. After all, John Lonsdale showed that the past is often represented in the present to the extent that "a society dominated by recollections of the resistances" takes a retrospective glance at the past to draw inspiration for ongoing and future struggles.[3] This insistence has much relevance to the contemporary Niger Delta where, in the context of festering environmental, political and economic crises arising from over four decades of crude oil exploitation, several communities are re-engaging with and deploying age-long idioms of resistance in remarkable ways. Studying these types of 'mobilization from below,' that is, from the perspective of the indigenous population, offers a completely different perspective than sticking to the "grand" narratives prominent in colonial discourses. How, for instance, did different groups interpret and deploy a wide repertoire of oppositional discourses and actions against colonial domination? What are the paradigms of exploitation and neglect, and how have these historically fed into and created the nexus between colonial domination and local resistance in the delta region? What politico-security and economic logics stimulated colonial domination and in turn, shaped resistance in that region, past and present? What mediating roles did foreign capital play in defining the contours and texture of domination and resistance in the delta, and what elements of change and continuities were evident? If, momentarily, we conceptualize encounters of domination and resistance as reflecting a nonlinear power relationship between British colonial authorities and delta peoples, how was this nexus implicated in the different contestations for power and authority? How did the relationship between the colonial state and colonial business shape the dynamics of domination and resistance and their various outcomes? In the final analysis, what linkages are there in the historiographies of domination and resistance in the contemporary Niger Delta?

This chapter is divided into three parts. The section that follows identifies and situates the key conceptual and theoretical issues germane to understanding the etiology of domination and resistance and a brief exposition on substantive methodological implications. Next, the chapter interrogates the different facets and manifestations of domination in the delta region and how these have conditioned the construction and attendant paradoxes of Pax-Britannica in that region. Finally, the discourse on resistance is introduced as a colonial anti-thesis deployed by coastal peoples in a variety of violent and nonviolent ways to challenge, negate and transcend the authoritarian excesses of colonial rule and what these portend for the texture and expressions of domination and resistance in the contemporary Niger Delta.

DOMINATION AND RESISTANCE: FRAMEWORK FOR INTERROGATING COLONIAL ENCOUNTERS

In theorizing domination and resistance, there is a need to situate the discussion within the wider debate on how local peoples reacted to the earliest contacts with and intrusions from outside, either by state or non-state forces. One strand argues that local peoples faced with this circumstance are prone to rebel as their traditional (or moral) economic security is weakened or destroyed. Beyond this orthodoxy, the rational-actor perspective[4] demonstrates that grievance and rebellions are dictated by perceptions of selective benefits or costs by local people. Whereas the former is the context that James Scott argued that peasant rebellions often occurred in situations in which peasants perceived a threat to what they regard as their rights to subsistence, the latter is represented by Popkin's view that local rebellions occur as a result of the divergent opportunities offered by modernization than owing to threats to local subsistence. Newbury proposed a mid-range view in which "conditions for revolution were created by long-term social transformations that substantially altered the position of subordinate groups in the context of the imposition of state power."[5]

Conceptually, "domination" and "resistance" are flip sides of the same coin in an environment where power relationships are uneven between the colonizer and the colonized. Thus, "domination" is the condition of having control or power over an entity, peoples or things by placing them in a position of subordination. On the other hand, resistance connotes a complexity of actions—sometimes even inactions—designed to blunt, retard, undermine, oppose, transcend or even escape a situation or condition perceived to have adverse effects.

It appears that some of the most blatant experiences of domination and resistance are found in—but not exclusive to—colonial relationships. In *Escape from Domination in Africa,* Bruce Baker gave some insights into the complexities of resistance and the domination impulses producing them.[6] Depending on the prevailing circumstances, domination and resistance can manifest in violent and nonviolent forms. In the colonial context, domination derives from the disposition of the colonial state to deploy the instrumentality of coercion against weaker entities. This pattern, which ran in varying scales (and duration) throughout British imperial presence in the coastal delta, shall be discussed in a subsequent section. Suffice that colonial domination is revealed, directly and indirectly, by and in the types of policies conceived and ferociously pursued by the different arms of British colonial administration, especially the civil service, police and the army. It involved frequent military actions sometimes at the slightest provocation, scorched earth, extraction of excessive reparation from the people postbellum, the arrest and exiling of ring leaders, heavy-handed policing and the maintenance of garrisons, all culminating in the imposition of pliant local political figures as titular traditional authorities. The array of local resistances, on the other hand, included small- or large-scale insurgencies, picketing, boycotts, banditry, rioting and armed rebellions. Whereas resistance of the violent mode is only the extreme response to domination, intervals of violent resistance are punctuated by pockets of intermittent nonviolent actions that help oppressed peoples to evade and or cope with excesses of their oppressors.

Even though violent forms of domination and resistance are themes privileged in the literature, the sublime nature of nonviolent gestures make them difficult or even impossible to curtail by the authorities. Furthermore, the very nature of nonviolent actions affords the local population many opportunities to undermine colonial rule without detection. Baker rightly insists that "[Just] as what is not said is often overlooked, though vital, in textual analysis, so what is not done . . . is often overlooked and yet vital in political analysis." Precisely for this reason, he further states, "[Between] the source of power and the individual there are, as it were, voltage loss, cable breakage, short-circuiting, and domestic disconnection reducing or interfering with the supply."[7] Because violent resistance tends to make the headlines, and is sometimes costlier in terms of human and material losses, nonviolent strategies employed to cushion domination generally escapes observation and suppression by those in authority. If the coercive and violent facets of domination and resistance are excluded from the discourse of colonial and post-colonial rule in Africa, as he rightly claims, we are likely to harvest a richer variety of alternative perspectives on how disadvantaged groups avoided or blunted the authoritarian disposition of colonial regimes using non-voting, non-rebellion, non-membership, non-conformity, non-payment of taxes etc. For Baker, this "escape from domination" could be the "product of the exigencies of particular times, social conditions and types of regimes" as their erosive and delegitimizing capabilities frustrate "the best laid plans of rulers for improving their authority and support."[8]

Disengagement, in turn, becomes the preferred option because of its remarkably ubiquitous nature: "[When] individuals or groups attempt to detach themselves from the unacceptable dominance of individual rulers and authoritarian systems" by ignoring, evading, mitigating or warding off (but not changing or destroying) the government "ceases to have legitimacy in their eyes."[9] Invariably, because of the ruthlessness that usually attended colonial rule, and the danger that lurked against resistances that involved "direct, often violent, confrontations between the wielders of power and dissident groups,"[10] the opposition is better rewarded by adopting avoidance rather than confrontation as the preferred method for expressing disaffection. Far from being isolated and episodic, therefore, avoidance is revealed by the impressive ways in which local communities mobilized and deployed them in the delta region during the colonial period.

The key insight from the above exposition is that even in its most intrusive form, colonial rule never completely penetrated every facet and inner recess of colonial societies. For instance, the manner in which the writ of the colonial state was deployed in colonial Africa merely secured limited presence for the authorities within the public sphere, whereas a range of traditional authorities, kinship and ethnic associations, as well as fraternities, dominated other equally important governance spaces.[11] It is precisely for this reason that any understanding of the nexus between domination and resistance must attend to the ecological context in which they are expressed and the likely limitations. It is often said that the way of life of a people is deeply etched in their geography and culture. This is even truer in the Niger Delta with its unique hydro-ecology that made it possible for different groups to adapt resistance. Thus, even if we take for granted many of the factors that make domination possible, much would still depend on the vagaries of geography—that is, whether zones of domination and resistance are mountainous, water-logged or plain savannah. Such physical relief will also determine the types of resistance options people are able and willing to deploy. The countryside, according to Desai and Eckstein is the

> arena of insurgency. It provides the ideal setting for collective political violence. Communications are poor and networks of transportation underdeveloped. Irregular terrain provides hideouts for dissident figures and great obstacles to authoritative forces, police as well as military . . . jungles, mountains, and peasant villages probably are not only boring, but mysteriously perilous.[12]

In *Anatomy of Rebellion,* Welch showed that rebellions are most likely to occur at the intersections of geography, history and social realities.[13] In other words, under conditions of geographical marginality, the type that is widespread in the delta region, past and present, "the writ of government receives little attention" just as its inefficiencies encourage "oppressive tenancy". To sufficiently account for why delta communities were able to keep Europeans at bay for almost four centuries, this fact of the advantages or otherwise of geography cannot be discounted. Desai and Eckstein have admitted that "the natural habitats of rural insurgents are also those of bandits: mountains, trackless plains, marshlands, and estuaries with their labyrinths of waterways. There they develop skills that are invaluable in insurgency operations."[14]

These perspectives would be incomplete without bringing in the role of culture, especially in the incubation of resistance. It is "essential to explore historical continuity," especially how the past and present factors interact [in] explaining domination and resistance in different settings.[15] The landscape of culture allows different societies to develop a unique constellation of norms, practices and institutions that would, in turn, help to shape what they considered important enough to fight for and the resources to commit to each emancipatory project. Due to the centrality of culture, aggressive behavior increases among (a) people with a strong belief in the efficacy of collective violence, (b) people exposed to social norms that facilitate and motivate compliance with the use of force and (c) people who regard aggressive behavior as normatively justifiable because of ideological orientation or perception of alienation from the economic opportunities and political systems.[16]

Our analysis thus far raises a number of methodological concerns, three of which have been acknowledged by Terrence Ranger: (1) the fact that scholarship has not insufficiently engaged with the modes of production historiography as a first crucial step in gaining better understanding of the social base of resistance in any society, (2) the disposition that is widely found in the study of resistance towards elitism and concomitant overshadowing of the role of the under-classes in the resistance struggles and (3) the admonition that scholarship may never truly grasp the character and meanings of resistance to different peoples as long as it fails to recognize the multiplicity of interests within African societies and how they interact with external forces.[17] These are partly why narratives suggesting that traditional elites and peoples are locked in the same resistance, or that elite consciousness dovetailed with that of the people, are undeserving—but unfortunately persisting—fallacies. Clearly, nothing can be more mistaken than assuming that the elites and peoples of the coastal delta spoke and acted with one accord. In the same regard, the discourse of resistance must get over what has become a pathological fixation with mass protest movements as if these were the most and only important genres of resistance. After all, the evidence is overwhelming that a whole range of localized oppositions existed; peasant revolts to tax evasion, refusal to perform forced labor, desertion, sabotage, social banditry, maroonage, withdrawal and other day-to-day acts.[18]

In the urgent task of restoring the "currency of the African past" therefore, we must return to Nwaubani's proposal for scholarship to return to "counter-history," the type that first-generation historians of Nigeria's colonial experiences deployed to debunk dominant Eurocentric perspectives on domination and resistance in the delta and elsewhere.[19] By engaging this mode of scholarship, current reinterpretations of African historiography are better able to illuminate the internal dynamics of societies and show that they can also be "active and autonomous participants, shaping and responding to events in the African environment."[20] If the historiography of domination and resistance in the Niger Delta contemplated here brings the people back in, then it would achieved its primary goal.

'STRUGGLING WITH THE HYDRA': COLONIAL RULE AND THE PARADOXES OF DOMINATION

One of the leading English historians of the African pre-colonial experience, Margery Perham, used the imagery of one out of the twelve labors of the Greek Hercules to describe the complications that attended British conquest of the peoples of the Oil Rivers and the hinterlands of contemporary Nigeria.[21] In that task, Hercules' frustration stemmed from having to battle a nine-headed serpent called Hydra, which grew two more heads each time one of its heads was cut off. This analogy is instructive, not just for accounting for the paradoxes of domination and resistance in the delta region in the colonial era, but also for grappling with the problems of governance and insecurity in the contemporary Niger Delta, which for the most part seems to defy solutions.

Following on the heels of the Portuguese, Dutch and the French, the British arrived at the west coast of Africa around 1533. Prior to their arrival, historical accounts indicated that various communities along that coastal strip established functioning political and economic relationships between and among themselves and with the hinterlands. Archeological records also showed that by the time nascent European colonial incursion commenced, it relied essentially on what was already on ground to survive and thrive. The historiography of the delta region must therefore go further than those defined by contact with Europeans, per se, to acknowledge the fact that state building and consolidation were already apace and did not derive essentially or solely from trans-Atlantic slave trade.

Despite festering disagreements, scholarship converges on at least one point about the traditions of origin of the different communities that make up the delta: whereas small settlements existed but far apart, many of them resulted from hundreds of years of migrations from different locations. Invariably, these waves of migration and settlement triggered an interface of collaboration and discord as different groups jostled for advantages, especially that of gaining access into and control of scarce habitable land, markets, trade routes and fishing waters. Prominent city-states therefore flourished in the delta region long before the advent of Europeans, especially in places such as Ebrohemie, Warri and Sapele in the western delta; Brass, Twan, Akassa, Ogbolomabiri and Bassambiri in the Ijaw country of the central delta; as well as in the Kalabari kingdoms of Buguma, Abonema and Bakana.[22] Expectedly, however, the waves of migration etched an indelible mark on the delta region as reflected in the divergence of its ethno-linguistic groupings. Despite its relatively small size and the constraints imposed by a difficult mangrove ecology, Nigeria's delta region is home to about three dozen ethnic groups, the most prominent being the Ijaw, Itsekiri, Urhobo, Isoko, Andoni, Ibibio, Efik, Kalabari, Ogoni, Okrika and the Ikwerre.

As previously mentioned, small fishing settlements and villages gradually transformed into popular trading posts and city-states[23]; the network of coastal communities began to cohabit, from time to time in less than affectionate ways.[24] This was accentuated by the imperative of gaining safer and secured access to the coast to gain maximum advantage from the nascent trans-Atlantic trade. This was the context in which Alagoa debunked "the creative effects of the European traders" on commerce and political developments in the delta as an exaggeration given that this new "overseas trade was grafted on to an earlier system of long-distance trade within the Niger Delta."[25] His point, quite instructively, is that both factors must be considered in equal measures: earlier long-distance trade provided the stimulus to the formation of state systems, whereas the "overseas trade merely altered the nature and dimensions of the trade within the Niger Delta and accelerated changes already begun by the internal long-distance trade."

With the points about the genealogy of coastal city-states settled, what is left is to aver that trans-Atlantic trade produced different patterns of domination, and invariably, also resistance. To capture this requires a delineation of three distinctive but linked eras in the contact between Europeans and inhabitants of the delta region. Roughly, these could be divided into (i) the period before 1807 when the trans-Atlantic slave trade was abolished, (ii) from the commencement of 'legitimate' trade in oil palm after 1807 to 1885 when the major European powers met in Berlin to arbitrarily partition Africa and (iii) from the post-1885 years onward when Britain imposed formal colonial rule.[26] Each of the phases coincided with different modes of accumulation in which the delta region played major roles: as the largest slave market in Africa, it accounted for more than half of the estimated 10 million slaves exported to the Americas during the first epoch.[27] It was also the largest market for the oil palm that lubricated the Industrial Revolution in Britain. More recently, since 1958 when crude oil was discovered in large quantities, the region accounts for more than 75% of Nigeria's foreign exchange earnings.

In the transition from one phase to another, however, there were striking elements of continuity rather than differences in the patterns of domination. For instance, the transition from slave to palm oil economy was marked by unpleasant frictions: between and among entrepreneurial coastal chiefs, on one hand, and major palm producers in the hinterlands, on the other, as well as that between the coastal chiefs and European supercargoes. In the scramble to control palm oil markets and trade routes, trade-related conflicts became frequent and dangerous throughout the 19th century.[28] By the late 1850s, significant changes occurred in the pattern of trade across the region largely owing to the establishment of fixed European premises ashore and the amalgamation of businesses formerly run by individuals into bigger limited liability conglomerates. In 1866, for instance, 17 European firms operated on the Bight of Biafra. By the 1870s, 279 Europeans were living and trading in Bonny, 270 in Elem Kalabari and 100 in Nembe, the three key oil palm markets. The capital base of British companies also increased from £243,000 in 1830 to £492,000 in the 1840s.[29] This heavy commercial presence also had at its disposal a strong British naval squadron costing about £706,454 per year to maintain.[30]

About the first quarter of the 19th century, coastal city-states began to turn into the ground zero of the lucrative West African (and global) palm oil trade.[31] However, behind this trade was concealed other items such as firearms and liquor, imported—but often smuggled—into the region. Kenneth Dike had complained that very little of permanent value came into West Africa from 400 years of trade with Europe in return for the superior labor force, the palm oil, ivory, timber, and gold which fed and buttressed the rising industrialism; they received the worst type of fake gin and meretricious articles. The abundance of firearms gradually ensured that "when hostilities break out they will be equally bloody to the natives, and injurious, if not dangerous, to the Europeans."[32] It was also fortuitous that the deepening European presence expanded the scope and context of social combustion during that period. Those conflicts often reached their peaks during difficult economic situations when fluctuating prices meant lower profits, dwindling market shares and uncertainties between commercial partners. This was exactly what happened in 1846 when King Pepple of Bonny launched an unpremeditated but successful attack on Andoni and annexed it to secure direct access to the Ibo oil producers in the interiors.[33]

Invariably, the delta region became the theatre of ferocious attempts to coerce, subdue and dominate local groups. By the 1850s, much of the initial skepticisms about the cost and benefit of greater insertion into and participation in the affairs of the delta disappeared among colonial officials replaced by a growing recognition that having a foothold in the delta was pivotal to the pursuit of British imperial ambition in that part of Africa. From then onward, increased consular presence began to confront and subdue, often with a fight, coastal potentates whose activities were deemed "inimical" and "subversive." The dawn of what turned out to be one of the most frightening eras of British gunboats commenced in full throttle. It was during this new phase that the modern variants of domination began to assume their forms, especially around the colonial state, colonial capital and the colonized peoples. Olorode reckoned that the "relationships of trade and exchange were not only exploitative and unequal as they have continued to be . . . they are built and maintained by intrigue, perfidy and armed violence directed against the indigenous peoples."[34] This, in turn, conditioned the behavior and responses of the "neo-colonial State and foreign-dominated capital to the resistance of the victims of neo-colonial plunder."[35]

It is no longer in contention that commerce had been the driving force behind European penetration of Africa and since then, the centerpiece of Euro-African relationships. One of the vocal ideologues of British (European) imperialism, E. D. Morel, unapologetically insisted that: "Commerce took us to West Africa; commerce keeps and will keep us in West Africa. It is the *fons et origo* of our presence in West Africa."[36] But then, commerce could neither stand alone nor in the least provide sufficient basis for the pursuit of British colonial expansion in Nigeria. It required far more concrete, intrusive and direct interventions that were political and military in nature to accomplish Pax Britannica within the territory. The first step was the proclamation of the Oil Rivers Protectorate (ORP) in 1891,[37] which later merged with the areas formally under the jurisdiction of the

Royal Niger Company, to form the Niger Coast Protectorate two years later. By January 1908, the Bank of British West Africa Limited applied to the Under-secretary of State for the Colonies for license to commence banking operations at Forcados and Calabar (CO520/73/3448).

The trajectory of British military (and political) presence in West Africa had received keen attention in historical research,[38] especially from the beginning of the early 19th century when its naval squadron was actively involved in antislavery campaigns, and later when several Constabulary Forces were established to discharge police and military duties in aid of commerce. Indeed, as trade expanded, the need for a succession of military expeditions and conquests grew not just in the Niger Delta but also further inland among the Yoruba, Igbo, Efik and the Ibibio. In these and a number of other cases, the rationale behind British military expeditions often ranged from the intangible to the absurd. In Ijonema, a community about three miles from Amie (headquarters of the Kwale District), Major Maclean led a force armed with a maxim gun to shell the town in October 1905, simply because a native court messenger sent by District Commissioner Davidson to serve a court summon was flogged.[39] By the time the soldiers arrived, inhabitants of the town had vacated but hoisted two white flags and left 26 calabashes of palm oil, 4 goats and 300 tubers of yam, suing for peace. Despite this peace overture, the entire town was razed by the invading army who went farther in hot pursuit of the fleeing people as far as neighboring towns. By the time the people capitulated, stiff penalties were imposed, including 100 goats, 20,000 tubers of yam and a commitment to undertake the construction of such roads as the political officer might order. The alleged ringleaders were also taken to Sapele as hostages to ensure the payment of the prohibitive fines. The soldiers also remained long after the cessation of hostilities ruthlessly exploiting local labor to construct a military garrison.[40] A similar situation arose during the Qua Iboe expedition of 1909 when a detachment of 6 European officers and 100 rank-and-file African soldiers drawn from the Royal West African Frontier Forces burnt the town into "a heap of blackened ruins" right up to the borders of the town. One of the officers who participated boastfully recounted how he single-handedly killed between 50 to 60 natives, including innocent citizens.[41]

Because they represented that face of British colonial administration, the police and military institutions shared common attributes of notoriety such as the Glover Irregulars in Lagos and the Oil Rivers Irregular, which formed the nucleus of the standing army of the Niger Coast Protectorate established in 1893. For instance, the rank-and-file recruits were individuals of questionable backgrounds: bandits, thugs, social miscreants and convicted rogues. There was also an unwritten rule that in the event of a breach of the peace, only non-natives who are likely to be less sympathetic to local people and their grievances are to be deployed, a practice that has continued to date. Thus, the disposition of the colonial government is to deploy a mercenary police system "whose personnel are alien to, alienated from and hostile to, the population among whom they are deployed."[42] Finally, they were essentially also veritable instruments for maintaining the power and authority of the colonial elites. In short, the "colonial police forces were organized and oriented to behave as occupation forces—ruthless, brutal, corrupt, dishonest and prone to brutalizing the colonized peoples and vandalizing their properties."[43] It is striking how these early compositions and behaviors continue to define the outlook of policing and soldiering in post-colonial Nigeria to date. Such were the imperfections and perversions that marked colonial dominations and British rule in the Niger Delta and the modalities of resistance that is considered next.

RESISTANCE AS DISCOURSE OF ANTI-THESIS: LOCAL STRUGGLES IN THE COLONIAL DELTA

In rereading the historiography of local responses to colonial rule among delta communities, it is striking how much they shared of the "discourse of anti-thesis" to domination. By this, I refer to the robust variety of locally contrived options they deploy, embedded as they were in shared sociopolitical, cultural and environmental conditions against the authoritarian excesses associated with colonial rule. As a discourse of anti-thesis to colonial domination, resistance evolved as one in a series of attempts to challenge hegemony building by the nascent colonial state. In *When Citizens Revolt,* Okonta showed how one of the most vociferous ethnic groups in the Niger Delta, the Ogoni, made sense of and negated the power and authority of British colonial officials at different times using "processes of resistance, accommodation and collaboration."[44] Within this logic, resistance is interpreted as a kind of revolt against subjectivization and the erosion of previously held citizenship. The idiom of relationship with and reactions to colonial rule represented a distinctive genre of social and political actions better studied from below by bringing the people back in than following "grand narratives" constructed from above.[45]

Beyond the reflections of historians of the Niger Delta, what is left is to respond to the following posers: how, for instance, did different coastal communities respond to alien domination as occasioned by the establishment of British colonial rule? What responses did the communities mobilize against external pressures? What factors and exegeses informed those choices, and how have they been understood in the mobilization of resistance? On what major issues, or concerns, did coastal communities mobilize resistances, and in what different directions did their actions take? The point had been made earlier that colonial rule never, at any given point, reached deep down enough into most African societies. In what ways, and with what resources, did coastal communities secure spaces of traditional and cultural exclusivity that became impregnable to colonial rule? In what variety of subtle but critical ways did the groups avoid or blunt the authoritarian pretences and excesses of the colonial regime? What local/traditional/cultural impulses gave voice and amplification to resistance? What types of resources, human and material, were often mobilized towards resistance? Apart from violent forms of resistance that shared strikingly similar characteristics, what other forms of nonviolent resistances were deployed by the peoples? Last, are there gender and class dimensions to the resistance of coastal peoples?

Realistically, only a few tentative answers can be accommodated here. Even then, my contentions and conclusions are based only on a few of the major pacifications by the British in the Niger Delta: against Kings Jaja of Opobo, Nana of Itsekiri and Koko of Nembe Brass, in 1881, 1894 and 1895, respectively. Detailed accounts of the circumstances that led to these punitive expeditions are widely available in the public domain, so my intention here is simply to distil and synthesize them and set the tone for understanding the characteristics of the resistance underway today. First, none of the three cases involved spontaneous action by the parties, i.e., the British authorities and the three kingdoms. Although different coastal polities initially resisted European incursions, they were obviously also not terribly averse to political relations, which they believed should be conducted among "equals" in a congenial and respectful manner. It was only when the overzealous colonial representatives began to meddle in and take sides with commercial interests, and went further by using local resources to undermine the authority of the kings, that relationships became troubled.

Indeed, at different points when things went well, each of the three kings that eventually fought those epic battles maintained cordial relationship with the British. King Jaja, for instance, availed his soldiers in the service of the British authorities as it struggled to quell the Ashanti Uprising of 1873–1874, a gesture that earned him a sword of honor from Her Majesty's Government. It was such cordiality that also initially marked the relationship between Nana and the British leading to his appointment as Governor of the Oil Rivers in 1884. In the three examples, however, the bone of contention was always about commercial and political rights and privileges accruable either to the kings or to the British. Whereas the kings believed they were entitled to charge a *comey,* a form of sales tax payable within their jurisdictions, the British officials accused each of them of turning into serial monopolists whose activities threatened "free trade." Thus, even as Vice-Consul Henry Johnston described King Jaja as "one of the most grasping, unscrupulous, and over-bearing of the mushroom kings who ever attempted to throttle the growing commerce of white men with the rich interior," the King expressed his exasperation thus: "it is with deep regret that I notice the distrust with which you look upon all the statements emanating from me, especially when I remember the confidence that your good brother and your predecessor have had in my word."[46]

The three kings harbored long and skin-deep mistrust and ill feelings against British officials for taking sides with European commercial enterprises, especially as they tried to get pass coastal middlemen. Despite several complaints by the Brass People of Nembe against the high-handedness and abuses of the Royal Niger Company (RNC) that made it virtually impossible to berth, set up market and erect stores without permits along and in traditional market sites. For instance, British consular authorities made no significant effort to discourage the monopolistic practices of the RNC—not to talk of sanctioning the company.[47] The Company, according to the people, adopted obnoxious and patently anti-people policies against Brassmen—including preventing them from collecting legitimate payments and debts from trading partners and seizing their "chop canoes" (i.e., canoes carrying food from the hinterland). This clearly represented an unconscionable contradiction, that the British would forcefully resist the monopoly by the Africans but condone its worst kinds by bigger commercial interests. There is no better account on how the activities of British commercial houses seriously undermined the livelihoods of coastal communities than the complaints by Brass people themselves that "We have suffered many hardships from the company's regulations, our people have been forced upon by the Niger Company' Launches, they have been fired upon from the Niger Company's Hulks, our canoes have been seized and goods taken, sometimes when engaged in what white men call smuggling, and sometimes when not." They were embittered that "[The] ill-treatment of the Niger Company is very bad . . . they fired, killed and plundered the fishermen and even the innocent women were caught, stripped naked, and painted with coal tar."[48] Finally, they justified their resolve to attack the offices of the RNC in Akassa, thus:

> A person who has been accustomed to eat [*sic*] bread and knows its sweetness from his youth to manhood happens to offered [*sic*] by another man to eat dust instead of bread. At the very time he hears the word dust he will go mad. Starvation brings a man to insanity." They concluded their letter by noting that: " . . . if part of the Niger River is not given to us the war with the Niger Company will not stop.[49]

For these legendary cases, and others of lesser note, a number of other factors seemed to have defined the texture of resistance and oppositional politics in the coastal delta. In no particular order of significance, three factors shaped most resistances: (1) the ecological backdrop within which they occurred, (2) the exegeses of cohabitation caused by the obtuse demands of trans-Atlantic commerce and (3) the types of cultural resources deployed by the communities. These factors are crucial to any understanding of the direction that resistance took, and continue to take, in contemporary Niger Delta.[50]

To give a sense of the ecological diversity of the Niger Delta, the region is at the point where the Rivers Niger and Benue empty their waters into the Atlantic Ocean. It covers an area approximately about 70,000 square kilometers, making it one of the world's largest wetlands. Despite its relatively small size and asphyxiating mangrove ecology, the Niger Delta is home to about three dozen groups known to have started as small fishing villages, metamorphosed into different sizes of city-states, and ossified into ethnic groupings. No wonder the region captured the attention of the earliest European visitors. Mary Kingsley, for instance, described it as one of the great swamp regions in the world in "its intensity and gloom, it has a grandeur equal to that of the Himalayas." Talbot called it the "region of somber mangrove," whereas John Hubbard referred to it morbidly as the "white man's grave," in apparent reference to the health incapacitations and deaths it caused among Europeans until the discovery of quinine and other antibiotics. Jones described the greater part of the delta as "a barren waste of mangrove forest and most of the inhabitants live scattered over it in small widely dispersed communities of fishermen."[51]

From the earliest period of European incursions through to the difficult establishment of formal colonial rule, the people of the Niger Delta maximized their deep knowledge of and familiarity with the coastal terrains to contend with and evade domination. Indeed, several colonial accounts openly expressed exasperation as to how coastal peoples successfully managed to use creeks as hideouts to evade the long arms of the punitive colonial state, or to inflict substantial damage on the edifices of the colonial state. Indeed, some of the epic battles in the region between coastal monarchies and the British were fought on such creeks and waterways, and often these involved setting up elaborate booby traps and maritime defense infrastructures along the waterways that left heavy casualties on the side of the invading armies. For this reason, local resistances were not limited to instances of full-blown wars but pockets of "strike and withdraw" tactics such as sabotage, banditry and sea piracy.[52] Because it was also possible to conceal many, if not, most of these local actions,[53] they exacted heavy tolls on colonial policing and intelligence services. The general implication of this situation, according to Flint, is that: "the attempt to get behind the middlemen on the Niger had reduced the banks of the river to anarchy." In the worst case scenario, such contestations brewed a groundswell in anti-British feelings among coastal middlemen.[54]

Frequently, irritations caused by the demands of expanding trade and other commercial activities made it imperative for coastal communities to be vigilant and alert to the prospects of small- or big-scale wars to enjoy stability and prosperity. From time to time, the looming prospect of war therefore frequently necessitated the acquisition of first-strike capabilities, the maintenance of credible defense systems against rival trading houses and the prevention of external attacks. In the process, different groups and communities armed themselves, but did so less intended to precipitate in conflict than to maintain credible deterrence and safeguard themselves from uncertainties. Those who claim, with pomp, that coastal groups were historically in perpetual breach of the peace betray a narrow appreciation of the complex and ever-changing imperatives of how trade frequently necessitated innovative self-help security arrangements suitable to the environment.

Resistance may have grown logically out of the circumstances of domination, but it also revealed the agility of coastal traders to anticipate and fully exploit different circumstances. In many instances, coastal peoples wittily employed a repertoire of strategies to deal with and extract favor from rival European commercial houses. To achieve this, coastal traders played the proverbial beautiful bride by pitching different merchants against one another, and in the process, maximized their individual and corporate benefit. This was the case with the complementary roles played by different supercargoes in terms of providing fiscal and military supports to the leadership of Anna Pepple House in October 1869 after Jaja was chosen as the leader. This support played a significant role in Jaja's eventual decision to establish the secessionist state of Opobo very close to the oil palm markets of the hinterlands. It is unlikely, as it were, that Jaja would have successfully stemmed the stiff oppositions from Bonny without the open support from a cross-section of supercargoes sympathetic to his ambition.[55] It was in a similar circumstance that key British commercial interests developed cold feet in the pursuit of efforts to break the monopoly of delta merchants after the catastrophic Niger Coast expedition of 1841. This was why, in the face of trenchant opposition by Brass people of Nembe who "coerced and armed" other smaller and weaker coastal neighbors to make trade more and more violent and difficult for Europeans, they continued to receive arms and munitions, as well as political support, from a segment of British traders, especially those originating from Liverpool.[56]

In a number of instances, elements within British commercial houses contributed, directly or indirectly, to the mobilization of local resistance and the success or failure of such ventures. In other cases, paradoxically, certain local groups openly cultivated and benefitted from European business and consular interests, thereby blunting local resistance. What these divergent trends would suggest, in part, is that we cannot always assume that groups with similar affiliations or in similar circumstances would necessarily pursue a shared vision of resistance (or, for that matter, of domination). As a matter of fact, given the manner that political and commercial relations were forged and pursued at that time, it would be unlikely that any convergence of interests would mature between and among the different coastal communities, or within the ranks of the supercargoes. To be able to maintain any competitive commercial edge over any length of time, it is necessary and expedient—indeed, a matter of enlightened self-interest—for European merchants who doubled as key arms dealers to court 'friendly' communities. Usually, it was on the basis of this type of double-edged relationship that the entrenchment of British colonial rule waxed or waned.

Given the abrasive divide and rule manner and the context in which colonial policies were pursued, it was not hard to find pliant local collaborators willing to work with British authorities to undermine local resistance. Again, in the three epic cases, the arrowheads of opposition against Jaja, Nana and Koko were easy to find given that their commercial and political successes already earned them enemies among weaker competitors. Because each of the kings rose to power against the backdrop of intense domestic political intrigues, any expectation that they would distance themselves from the controversial circumstances during their reigns was hopelessly mistaken. In the case of Nana, for instance, the British simply found and propped up Chief Dogho, his sworn enemy. In the case of King Koko of Nembe, the circumstance was quite different: all the chiefs and their constituencies were united against a common enemy, the Royal Niger Company, as demonstrated by the absolute secrecy and military precision with which they orchestrated and executed the attacks on the facilities of the company in 1895.

Whereas it developed out of the need to secure trade, one of the most institutionalized infrastructures of resistance against European commercial and consular presence in the delta was the Canoe House system. This system has received a lot of attention in historical research on the Niger Delta and has been the focus of a recent study seeking to give a more theoretically inclined explanation for its emergence and far-reaching significances.[57] In *The Small Brave City State,* Alagoa described this structure as "far from being primarily a commercial corporation. . . . It was rather an association for maintaining law and order in society—a form of social security and insurance."[58] Canoe Houses constituted a veritable defense mechanism against a wide range of threats: from bandits and pirate attacks to commercial warfare aimed at undermining each other; and finally, as a bulwark against European mercantile and political authorities. Deriving from this multiplicity—not necessarily the duality—of the role of Canoe Houses, Jones referred to them as "fighting corporations."[59] Whatever perspectives may have been offered about the Canoe House system as one of the most innovative trade infrastructures utilized by delta communities, they were quite an important competitive response to trade and the mobilization of resistance against adversaries; after all, to keep commerce running is tied to the maintenance of a credible military.

How the imperatives of commerce stimulated local resistance is shown further by the commitment different coastal city-states made concerning protection and security. There is no better indicator of this than the unending quest to acquire various types of firearms from friendly European merchant houses. Second, and against the backdrop of their superior firepower, the coastal communities were able to coerce and arm their neighbors to make business difficult for competitors. Because resistance neither occurred in militarized forms nor followed a unitary pattern,[60] sharp practices were often deployed by coastal middlemen in dealing with their European counterparts, regardless of whether such methods ended in disastrous consequences or retribution. One such method of trade manipulation routinely used was that of price fixing, for instance, that which took place between Kalabari canoe men and Aboh traders to regulate the prices paid for palm oil in 1905. In that instance, European traders received assistance from the District Commissioner to charge 56 "truculent" Kalabaris and 14 other African traders to the Native Authority Court at Oguta and imposed various fines on them.[61]

The widespread availability of firearms and ammunitions soon gained notoriety as one of the biggest threats to peace and security in the Niger Delta. Livingstone observed, with much disenchantment, that "English traders have furnished the rival chiefs with guns, shots, and powder, as required, until war materials are a drug" on the delta markets. This trend not only culminated into an unhealthy armament race but also "intensified the struggle and widened the area of conflict to the tribal interior."[62] No matter the dangers they ultimately posed, however, the possession of firearms and ammunitions conferred a lot of strategic advantages, not the least the advantage of first-strike capability.[63] It was only a short time before different coastal communities developed greater familiarity with and proficiency in the use of the numerous weapons and ammunitions obtained over years of overt and clandestine commercial relationships with European supercargoes. This was stated, perhaps in a slightly exaggerated overtone, in a correspondence by Captain W. E. Rudkin of the Royal Artillery to the Secretary to War Office in London over the Agbor disturbance (read: resistance). According to Ruskin, "[Everyman] from youth up carries a gun and knows how to use it." The salience of this, according to him, was that the people put up a "stout resistance" and "tenacity," which led to the death of 19 soldiers, 113 wounded on the side of the British forces and almost 200 within the community.[64] During the Niger expedition of 1841, also, 48 out of the 145 European participants reportedly died.[65] Another punitive expedition to Kwale country in 1904–1905 suffered a similar fate owing to the stout resistance mounted by the people and high casualties suffered by both sides. Again, a dispatch by the Officer Administering the Government of Lagos and Southern Nigeria to Sir Lyttleton at the Colonial Office in London in 1905 attested to this by noting that all three British officers were wounded, apart from another 14 and one dead.[66]

These instances have been highlighted, in part, to underscore that even though punitive expeditions against coastal communities were all too often heavy-handed and brash, they never really deterred the people from taking up arms when they deemed fit. They also demonstrate the dangers of underestimating both the amount of weapons at the disposal of the different delta communities and their resolve to 'fight to the finish.' In the case of Kwale, for instance, colonial dispatches recorded that the resistance was so swift and robust that the Commanding Officer had to urgently request for reinforcements. This explained why colonial authorities routinely insisted on comprehensive and unconditional disarmament, and the cashiering of weapons, after major hostilities. In the aftermath of one of the most prolonged and ruthless expeditions by British colonial authorities against the Aro people in 1902, archival records showed that it took an overwhelming British firepower (comprising of 87 officers, 1,550 men and 2,100 carriers) almost five months (from November 1, 1901 to March 3, 1902), and an unprecedented casualty rate on both sides, to subdue the people.[67] A raster even indicated that the British captured over 25,000 guns, mostly rifles and cap guns, as well as one 75 mm battery gun, one 7pr four-gun battery, 62 mm guns and ammunitions. It was on the account of the overwhelming military resources at the disposal of the Aro people, even though they are not located physically in the coastal delta to benefit directly from European arms trade, that the British promulgated the Firearms Proclamation of 1902 outlawing the possession of and trading in weapons and ammunitions.[68]

More sublime than geography and widespread availability of firearms was the potency of culture in the deployment of resistance to colonial rule. To properly situate the role of "culture"[69] in incubating and amplifying local resistance in the delta region is to return to the point made earlier that at no time did colonial intrusion and domination effectively reach every facet of sociocultural life throughout colonial Africa. What happened was that colonial rule actually created two organically detached spheres, or what Ekeh classified as the "two publics": the civic and primordial spaces.[70] Whereas colonial rule was alive and active in the civic sphere, it was not at all, or only partially visible, in the primordial space; within this later space, or pocket of spaces in physical or metaphysical senses, local resistances are hatched and amplified. The significance of this sphere is underscored by the fact that its members shared similar idioms of cultural, social and linguistic traditions that defined patterns of collective struggles they wished to embark on, and with what resources. When local resistances are incubated within such spaces, members are able to 'own' the basis for which they are called on to embark on a range of protests from theatrical satire and ridicule to impromptu gestures of contempt and mockery that flourished among local communities but often passed unnoticed by colonial officials. Finally, such spaces usually served as the realm of protection were harm avoidance counter-measures are incubated. Because resistances are first and foremost sanctioned by customs and traditions, they played a major part in defining what people would acquiesce to or vehemently resist.[71]

Still on the role of culture in the mobilization of resistance, the final issue relates to the kind of cultural impulses that inexorably gave voice and meaning to local resistance among coastal peoples and communities. It is important to bear in mind, first, that given the arbitrariness and authoritarian disposition of colonial authorities, the first line of response was usually for the affected community to sharpen their individual and collective reflexes and instincts for survival in the face of danger. In communities under colonial siege, mentally and physically, members are predisposed to strengthening or reclaiming culturally unique idioms of resistance by belonging to fraternities and brotherhoods such as the Ekpe Fraternity among the Efiks of Old Calabar that afforded them the best guarantees and protection against arbitrary colonial rule.[72] It is in a similar circumstance that the relevance of Kpone-Tonwe's detailed insights into the Yaa traditions as one of the many cultural undercurrents giving meaning and direction to past and recent expressions of Ogoni nationalism must be understood. In his account, the Yaa tradition "provided a means for the steady recruitment and training of Ogoni youth for the defense and preservation of their society."[73] Although Kpone-Tonwe feared that the tradition is becoming extinct, the last rites having been performed in 1983, this conclusion may have missed the sense of cultural revivalism that contributed to the resurgence of Ogoni nationalism from the 1990s onward.

This type of cultural revivalism may have survived and lingered in the popular consciousness of the people, even where the rituals and symbolisms associated with them may have atrophied for several decades. Kpone-Tonwe hinted discreetly at this: "the age-old environmental tradition of the secret societies provided the basis for popular awareness of the environmental campaign against Shell."[74] Given the sheer diversity of their traditional and cultural institutions, it is not unlikely that coastal communities in the delta region would draw on their past experiences, especially from the most difficult days of colonial rule, and now also, as demonstrated by the Ijaws invocation of the powers of their mythical god, Egbesu, to challenge the myriad developmental problems facing them in the context of the contradictions of oil-based accumulations. This, of course, is only a tentative—but informed—guess that deserves deeper theoretical scrutiny.

To conclude, a point about the nexus between colonial rule and ethnicity is due. We have already shown the significance of the changing natures of trade and politics and how they influenced domination and resistance in the delta region. It is important to mention that the peculiarity of domination and resistance in the delta region may have provided the impetus for the kind of ethnically constructed mobilization and the tensions and violent conflicts they produced during and after colonial rule, and even in contemporary times. Throughout, British colonialism tinkered with and manipulated ethnic sentiments to great length, especially so in the delta region, which has become home to a plethora of ethnic and sub-ethnic groupings locked against one another in a variety of unaffectionate cohabitations. While it lasted, colonial rule favored and pitched some ethnic groups against one another, and such distrust and crises that grew out of that doused efforts to muster a united, pan-delta agenda and front against colonial rule. It is only appropriate to acknowledge that the different ethnicities and ethnic identities that matured across the delta region, and indeed across Africa, showcases how much ethnicity became a credible response to colonial, and post-colonial, rule.[75]

In the past, as well as in contemporary Africa, ethnic identities have endured precisely because they turned out to be a veritable resource for meeting and achieving politically defined ends. In many senses, as Okonta recently argued, the resilience of ethnicity invariably also marked very fundamentally the "story of the local people and the strategies of adaptation and resistance which they have devised and adopted."[76] Over the centuries, the Niger Delta has steadily grown into a site of stronger, more vociferous and more assertive identities with mobilization centered on the clamor for greater political and fiscal autonomies.[77] The failure to tame such radical persuasions seemed to have elevated violent, extra-legal means to a fad. What the future portends for the region, if the realities of the past and present times are considered, is frightening.

NOTES

1. Terrence Ranger, "The People in African Resistance: A Review," *Journal of Southern African Studies* 4, no. 1 (1977): 146.
2. K. O. Dike, *Trade and Politics in the Niger Delta, 1830–1885* (Oxford: The Clarendon Press, 1962); E. J. Alagoa, "Long Distance Trade and States in the Niger Delta," *Journal of African History* 11, no. 3 (1970): 319–329; E. J. Alagoa, *A History of the Niger Delta: An Historical Interpretation of Ijo Oral Tradition* (Ibadan, Nigeria: Ibadan University Press, 1972); J. C. Anene, "The Foundations of British Rule in Southern Nigeria, 1885–1891," *Journal of the Historical Society of Nigeria* 1, no. 4 (Dec. 1959); E. A. Ayandele, "External Influence on African Society," in *Africa in the Nineteenth and Twentieth Centuries,* ed. J. C. Anene and Godfrey Brown (New York: Humanities Press, 1972), 133–148; Obaro Ikime, *Merchant Prince of the Niger Delta* (London: Heinemann, 1968); Obaro Ikime, "Colonial Conquest and Resistance in Southern Nigeria," *Journal of the Historical Society of Nigeria* 6, no. 3 (1972): 251–270; Obaro Ikime, *The Fall of Nigeria: The British Conquest* (London: Heineman, 1977); Tekena Tamuno, *The Evolution of the Nigerian State: The Southern Phase, 1898–1914* (London: Longman, 1978).
3. Ranger, "The People in African Resistance," 126.
4. Mancur Olson, *The Logic of Collective Action: Public Goods and the Theory of Groups* (Cambridge, Mass.: Harvard University Press, 1979); Samuel Popkin, *Rational Peasants* (Berkeley: University of California Press, 1979).
5. Catharine M. Newbury, "Colonialism, Ethnicity, and Rural Political Protest: Rwanda and Zanzibar in Comparative Perspective," *Comparative Politics* 15, no. 3 (April 1973): 271–272.
6. Bruce Baker, *Escape From Domination in Africa: Political Disengagement and its Consequences* (Oxford: James Currey, 2000).
7. Ibid., 1.
8. Ibid., X–XI.
9. Ibid., 2.
10. Michael Adas, "From Avoidance to Confrontation: Peasant Protest in Precolonial and Colonial Southeast Asia," *Comparative Studies in Society and History* 23, no. 2 (April 1981): 217.
11. Baker, *Escape from Domination, 5.*
12. Raj Desai and Harry Eckstein, "Insurgency: The Transformation of Peasant Rebellion," *World Politics* 42, no. 4 (July 1990): 442.
13. Claude E. Welch, *Anatomy of Rebellion* (Albany: State University Press of New York, 1980), 2.

14. Desai and Eckstein, "Insurgency," 453.
15. Ibid., 445.
16. Steven E. Finkel and James B. Rule, "Relative Deprivation and Related Psychological Theories of Civil Violence: A Critical Review," *Research on Social Movements, Conflict and Change* 9 (1986), 62–63.
17. Ranger, "The People in African Resistance," 141–142.
18. Ibid., 135–137.
19. Ebere Nwaubani, "Kenneth Onwuka Dike, Trade and Politics, and the Restoration of the African in History," *History in Africa* 27 (2000): 237–238.
20. Ibid., 248.
21. A. E. Afigbo, "Britain and the Hydra in the Bight of Benin: Towards a History of the Abolition of the Internal Slave Trade in the Oil Rivers and Its Hinterland, C. 1885-C. 1943," *African Economic History* 31 (2003): 15.
22. E. J. Alagoa and A. Fombo, *A Chronicle of Grand Bonny* (Ibadan, Nigeria: Ibadan University Press, 1972).
23. E. J. Alagoa, "Oprozo and Early Trade on the Escravos: A Note on the Interpretation of the Oral Tradition of a Small Group," *Journal of the Historical Society of Nigeria* 1 (1969): 151–157.
24. For short accounts of the historical background to the Kalabari-Bonny conflict, circa 1790s, see Ume E. Kalu, *The Rise of British Colonialism in Southern Nigeria, 1700–1900* (New York: Exposition Press, 1980), 27–28.
25. E. J. Alagoa, "Long-Distance Trade and States in the Niger Delta," *The Journal of African History* 11, no. 3 (1970): 319.
26. It is important to note that matters relating to the delta region featured on the agenda of the major European powers that attended the infamous Berlin Conference of 1884–1885.
27. Kalu, *The Rise of British Colonialism,* 5.
28. Ibid., 16.
29. See W. E. Wariboko, "Social and Political Developments," in *Land and People of Nigeria: Rivers State,* ed. E. J. Alagoa, and Tekena N. Tamuno (Port Harcourt, Nigeria: Riverside Communications, 1989), 125–129.
30. Up to that period, as Hutchinson noted, this naval flotilla had claimed the lives of 166 officers and men and 104 invalidated. T. J. Hutchinson, *Narrative of the Niger, Tshadda and Benue Exploration* (London: Frank Cass Co. Limited, 1960), 246.
31. Dike, *Trade and Politics in the Niger Delta,* 42–50; John E. Flint, *Sir Goldie and the Making of Nigeria* (London: Oxford University Press, 1960), 10–11.
32. Dike, *Trade and Politics in the Niger Delta,* 107.
33. Ibid.
34. Omotoye Olorode, "Imperialism, Neo-colonialism and the Extractive Industries in Nigeria," in *Ken Saro-Wiwa and the Crises of the Nigerian State,* ed. Omotoye Olorode et al. (A CDHR Special Publication, 1998), 3–4.
35. Ibid., 16.
36. J. C. Anene, "The Foundations of British Rule in Southern Nigeria, 1885–1891," *Journal of the Historical Society of Nigeria* 1, no. 4 (Dec. 1959): 319.
37. At the time it was created, the Oil Rivers Protectorate was equal to, if not more viable than, the districts administered by the Governors if Sierra Leone, Lagos and the Gold Coast, prompting Major MacDonald, the British Consul on that coast, to request better remuneration for colonial officials posted to the areas. See FO403/133: Memorandum by Major MacDonald on the Oil Rivers Establishment, FO December 23, 1890.
38. For a very interesting account of the exploits of the RWAFF, see J. U. J. Asiegbu, *Nigeria and Its British Invaders, 1851–1920: A Thematic Documentary History* (New York: NOK Publishers International, 1984), XXV–XXVIII.

39. The contingent had loaned a millimeter gun from Calabar for just that purpose, making its intentions very clear.
40. CO520/32: Southern Nigeria. 1905, vol. 4, no. 38260 of 10/27/1905-Report by the acting High Commissioner, southern Nigeria to Lyttleton, SSCO, 11/18/1905.
41. For letter by R. Smith to M'Keown, Qua Iboe Mission, Belfast, 10/6/1909, See CO520/87/24005: Qua Iboe Mission: Punitive Expedition at Enen on Qua Iboe River.
42. Etannibi E. Alemika, "Colonialism, State and Policing in Nigeria," *Crime, Law and Social Change* 20, no. 3 (1993): 202.
43. Ibid., 204.
44. Ike Okonta, *When Citizens Revolt: Nigerian Elites, Bid Oil and the Ogoni Struggle for Self-Determination* (Trenton, N.J.: Africa World Press, Inc., 2008), 16.
45. Ibid., 24.
46. See Johnston to Foreign Office, FO84/1828. The harassment of King Jaja, as the others, should by no means be construed as representative of the stance of all segments of British colonial administration. Indeed, the report of an investigation by Commander Hammick to Commodore Sir F. Richards dated November 21, 1881, had argued that Jaja's case was a native, not trade, issue. Even Lord Salisbury who presided over the Colonial Office in London was recorded to have described the eventual deposition and exile of king Jaja as "kidnapping"—even though he did nothing to reverse the situation.
47. For emphasis, some of the European businesses and colonial officials on the coast were also very suspicious of and distressed by the activities of the RNC. Indeed, many of the British naval squadrons and administrators looked down on the 40 or so European employees of the company and treated them with contempt by virtue of their social statuses and lack of formal education.
48. See also FO403/215: Kings and Chiefs of Brass to MacDonald, 4/2/1895.
49. Thomas Hodgkin, *Nigerian Perspectives: A Historical Anthology* (London: Oxford University Press, 1975), 312.
50. There is still a controversy over the number of ecological zones within the region. See, for instance, The World Bank (1995), who identified five: coastal barrier islands, mangroves, freshwaters, swamp forests, and the lowland rain forest.
51. G. I. Jones, *The Trading States of the Oil Rivers: A Study of Political Development in Eastern Nigeria* (London: James Currey, 1963).
52. Ukoha Ukiwo, "From 'Pirates' to 'Militants': A Historical Perspective on Anti-State and Anti-Oil Company Mobilisation Among Ijaw of Warri, Western Niger Delta," *African Affairs* 106, no. 425 (2007): 587–610.
53. It was a big surprise to the Royal Niger Company, for instance, that the people of Nembe Brass would plan such an elaborate attack on its stores in Akassa in 1895 without any notice. It explained why Sir Goldie was so piqued that he was bent on extracting substantial war reparations from the people and to retrieve their weapons.
54. Flint, *Sir Goldie and the Making of Nigeria*, 18–23.
55. Alagoa and Fombo, *A Chronicle of Grand Bonny*, 73.
56. Flint, *Sir Goldie and the Making of Nigeria*, 20.
57. Nini Wariboko, "A Theory of Canoe House Corporation," *African Economic History* 26 (1998): 141–172.
58. Joe Ebiegberi Alagoa, *The Small Brave City State: History of Nembe-Brass in the Niger Delta* (Ibadan, Nigeria: University Press, 1964), 18.
59. Jones, "The Trading States"; Nimi Wariboko, "A Theory of the Canoe House Corporation," 141–172.

60. Colin Newbury, "Trade and Technology in West Africa: the Case of the Niger Company, 1900–1920," *The Journal of African History* 19, no. 4 (1978): 551–575.
61. Ibid., 560.
62. Dike, *Trade and Politics,* 194–195.
63. A. F. C. Ryder, "Dutch Trade on the Niger Coast During the 17th Century," *Journal of the Historical Society of Nigeria* 3, no. 2 (1965).
64. CO520/54: 15523-Southern Nigeria, vol. 12, 1907.
65. Michael Crowther, *The Story of Nigeria* (London: Faber and Faber, 1962), 9–14.
66. CO520/32: Southern Nigeria, vol. 4, 1905.
67. CO520/14: Southern Nigeria, vol. 2, 1902.
68. CO520/14: 18682 of May 12, 1902.
69. Here, I take culture simplistically to mean a framework or map of meanings and symbols that guide the behavior, attitude and disposition of a defined group of people living apart or in close proximity and sharing in a concrete or metaphysical sense, the same worldview.
70. Peter Ekeh, "Colonialism and the Two Publics in Africa: A Theoretical Statement," *Comparative Studies in Society and History* 17, no. 1 (1975): 91–112.
71. Adas, "From Avoidance to Cooperation," 227.
72. Nwaubani, "Kenneth Onwuka Dike," 235–236.
73. Sonpie Kpone-Tonwe, "Leadership Training in Precolonial Nigeria: The Yaa Tradition in Ogoni," *The International Journal of African Historical Studies* 34, no. 2 (2001): 385–403.
74. Ibid., 403.
75. Ukoha Ukiwo, "On the Study of Ethnicity in Nigeria" (CRISE Working Paper Series No. 12, Queen Elizabeth House, University of Oxford, June 2005), 9.
76. Okonta, "When Citizens Revolt," 7.
77. E. J. Alagoa and Tekena N. Tamuno, eds., *Land and People of Nigeria: Rivers State*, 6.

11 Education Transformations and Traditional Medicine in Nsukka, Nigeria

Jamaine Abidogun

1. What do you want to achieve with your secondary education?

Student: To be a medical doctor.
Interviewer: But if you only have secondary education, how does it help you?
Student: It helps me in some aspects, like for instance science, that is the main course.[1]

INTRODUCTION

What happens to indigenous or traditional medicine when Western education, in particular Western concepts of science and Christianity, dominate formal education in African societies? This study looks at the case of Nsukka Local Government Area or Northern Igbo societies located in Enugu State. The overall result is traditional medicine becomes viewed by Western educated Igbo at best as an "alternative" or "complimentary" medical source and at worst as "evil acts." It is important to consider the dual effect of Western Science and Christian doctrine. Christian doctrine provides a reason to reject the role of the dibia or herbalist along with their medicinal abilities.[2] Western science provides the idea of "validation" through experimental or clinical tests with written proof. If the "medicine" was not proven to work through this method and written documentation provided, then it must not be reliable or it may even be dangerous. These two factors work together to marginalize, even negate, indigenous sciences. The specific outcomes on formal education participants are much more varied than at first anticipated. These influences are explored through the student and parent and adult participant voices in Nsukka LGA.

STUDY SITE AND PARTICIPANTS

Nsukka Northern Igbo culture was chosen as the core cultural context because regardless of whether the participant identified as Nsukka Igbo or not, all participants lived in Nsukka and therefore shared some common experiences from that particular society. The research used the participants' understanding of their culture and society (both home culture and resident culture) in identifying currently accepted norms, values and practices. This method also historically positioned the information within and around the lifetimes of the participants. The participants' understandings took precedence to ensure that the value structures of their perceptions were considered, with secondary readings referenced to provide a broad, general understanding of Igbo culture. Although 95% of students identified as Igbo, regional and ethnic differences as it related to various cultural norms, values and practices were considered within the analysis and interpretation of the data.

Nsukka LGA is located in Eastern Nigeria in the northern part of Enugu state. Nsukka LGA includes several small villages of which one is Alor Uno where both school sites for the research were located. The University of Nigeria is the largest employer in the formal salaried sector, although farming and petty trades are the two main occupations of Nsukka residents. Whereas valid current census information was unavailable for the area, a 1999 university survey indicated an estimated adult illiteracy rate at 36%. In this same survey approximately 28% of participants identified as traditional believers, 66% as Christian, 2% as Moslem and 4% as having no religious preference.[3] Nsukka is predominantly Igbo but shares borders with some Igala/Idoma villages. The above demographics indicate a comparatively homogeneous participant profile.

Extensive interviews were conducted during 2004 and 2005 in Nsukka LGA.[4] Secondary students made up the greatest number of interview participants. These students were from an area parochial school, Shalom Secondary School, and a government school, Alor Uno Secondary School. All student participants identified as Christian, and the majority identified as Igbo at both schools. Thirty students from each school were interviewed twice. The first interview consisted of collecting student demographic information and informing each participant of the purpose and goals of the research. Also the first interview helped to establish a positive rapport with each participant, with a goal to optimize the second interview results. The second interview consisted of a series of questions to assess cultural changes based on formal education experiences. Many of these questions asked participants to compare current cultural practices with past practices based on family histories. Students, parents and adults who had attended secondary school in the area, and teachers at the schools, were also included as participants in this study. In addition to a range of formally educated Nsukka LGA residents, a biochemist whose work is directly tied to the intersection of indigenous medical science and Western science provided an extended interview for purposes of this study.[5]

The guardian and other adult formally educated interviews provided more detailed information regarding their knowledge and experience with dibias and/or traditional herbalists in the Nsukka Igbo community. This provided validation by comparing what students reported as their parents' experiences to what parents and other adults self-reported. The terms "dibia" and "herbalist" were often interchanged by participants during interviews. It was commonly understood that a dibia was always considered an herbalist, whereas an herbalist may or may not be a dibia. Few participants made a clear distinction and connected both titles to the practice of traditional religion. Two origins of influence surfaced from the collective interviews: one—Christianity and two—Western Science Curriculum as introduced through a Western education model.

THEORY AND INTERPRETATION

Neo-colonial theory, critical pedagogy and transformational learning theory informed the interpretation of data. Neo-colonial theory offers a way to historically situate the research. Understanding the imposition of Western education, medical and religious institutions within Nigeria allows for a starting point that underlines the forced historical nature of changes in education, science, medicine and religion. It further illuminates the nature of internalization of Western concepts as they replaced indigenous forms of knowing and doing. Simultaneously it recognizes that portions of the Nigerian population were impacted to greater or lesser extent based on the degree of their direct involvement with these institutions (colonial and post-colonial). Thus neo-colonial theory takes into consideration historical and present inequities in power relations based on political, economic and social realities within the society. In particular it provides contextual knowledge needed to consider the continued influence of Western capital (material, knowledge, religion) in influencing power relations in Nsukka society.

Critical and transformative learning theories emphasize the necessity to place the subject at the center of the research, so that their voices are the litmus test to validate research conclusions. Critical theory provides a basic understanding of Western education constructs as status driven by a capitalist and individualist philosophy that culturally consisted of a male dominated hierarchy within a white, Western, privileged context.[6] Transformational learning theory builds on this basis by synthesizing neo-colonial theory with critical theory and makes indigenous peoples' voices the focal point of political, social, spiritual, cultural and pedagogical struggle and change resulting in varied transformations. This theory pulls from the colonial and neo-colonial experiences of indigenous peoples from around the world and aims to provide strategies for decolonization. Spirituality is one area frequently left out of critical theory arguments, as these well intentioned scholars tend to present Western secular thought on class, gender and race divisions. It is within the indigenous peoples' histories and cultures that the integration of spirituality becomes crucial to understanding their struggle in neo-colonial form as they seek to reconstruct and maintain their histories and cultures. Shilling insists that for teachers and students, "Learning must be seen as an essential component of decolonization . . . it is through the learning process that we may grow and heal our historical and contemporary wounds."[7] In this way the participants may reclaim the educational process and there can be reintegration of indigenous ways of thinking as a part of educational practice. Also, Dei speaks specifically to the experience of African peoples, stating, "students want to learn the local knowledge and Indigenous experiences, but they also want to come out of their formal schooling with degrees and diplomas that are 'transportable' in globalized [neo-colonial] contexts. For all [African] students struggling in such educational contexts, there have been some losses along the way or at best the parallel moving along of two fragmented knowledge streams."[8] Dei captures the conflict and cultural displacement created through participation in formal Western education models that continue to focus almost exclusively on Western sciences. Use of a Western education model in the Nsukka LGA has privileged Western content (science and religion) within Nsukka society. The result is active resistance to traditional religion and no formal acknowledgement of indigenous medicinal knowledge within the education curriculum. This creates a dangerous situation where indigenous medicinal knowledge may be viewed the same as quackery or the knowledge may even be lost as an increasing percentage of Nigerians enter and complete secondary education programs.

EDUCATION IN HISTORICAL CONTEXT

European contact had an extensive and lasting effect on Nigerian education. In 1842 the Wesleyan Methodist church established the first British missionary school. At the turn of the century England took formal control of the Northern and Southern Territories of Nigeria. By 1910 the first British run secular school was established.[9] The British curriculum was mandatory in any public school and was the norm for missionary schools. The British commonly utilized Christian missionaries to establish and run schools, as this proved a cost effective practice under colonialism. Wolf Roder commented on the result of this historical partnership: "What we tend to think of as 'Western civilization' was thus not presented as a rational-logical system and approach to life, but rather as a superior gift from a truer god."[10] It is vital to understand Christianity's impact on Western education within the Nigerian historical context as one of the main reasons why the inclusion of indigenous knowledge in the colonial and post-colonial education system remained negligible at best. Even though the Western sciences were gradually added to the curriculum, the partnership that established Western and Christian as one and the same did not lessen significantly over time.

British curriculum consisted of six years primary school, five years secondary and two years senior secondary. The senior secondary level awarded a certificate roughly equivalent to an associate degree.[11] A three year bachelor degree was available after 1948. It should be noted that outside of major urban areas in Southern and Southwestern Nigeria, most communities only supported primary schools well into the 1930s and more rural areas into the 1980s. The first university was established in Ibadan as an affiliate of the University of London. The purpose was to train medical students. The curriculum also included teacher training and some survey courses. No full medical degree was granted by the University of Ibadan without the student's compulsory attendance at the University of London until 1960.[12]

During the early 1950s the need for Afrocentric education began to be addressed as Nigeria looked toward independence. Nnamdi Azikwe, first governor general of Nigeria, remarked on the mono-cultural character of British and European curriculum and the need for inclusion of African content:

> Black Africa has no university. Black Africa has no intellectual centre where the raw materials of African humanity may be reshaped into leaders in all fields of human endeavor. Why should African youth depend upon Oxford, Cambridge, Harvard, Yale, the Sorbonne, Berlin, Heidelberg, for intellectual growth? These universities are mirrors which reflect their particular social idiosyncrasies . . . give the Renascent African a University . . . and this continent can become overnight a continent of Light.[13]

The need for Afrocentric education would eventually be acknowledged. The majority of European trained Nigerian elites learned detachment and disdain for indigenous knowledge, whereas Azikwe envisioned education for and about Africa. The curriculum was slow to change whether in primary and secondary education or in higher education. Micheal B. Adeyemi, professor of education, noted that Nigerian social studies focused on "rote memorization, moral education, dislike for indigenous culture, and respect for western education."[14] In 1960 Nigerian independence brought a renewed call for a nationalized and indigenous education system. The goal to nationalize was apparent in social studies curriculum. In 1969 Nigerian government officials met with the United States Agency for International Development (USAID) and Ford Foundation representatives to create an educational curriculum that would address both Nigerian nationalization and the multicultural reality of Nigerian society. This meeting was held at Aiyetoro Comprehensive High School. There were extensive changes made to the junior and senior secondary curriculum to include African, and in particular Nigeria, history and cultures. This extended to changes in the language curriculum where a mandate was established that all schools require language courses in the regional mother tongue language and at least one additional Nigerian language. English remained mandatory along with Western sciences within the secondary curriculum.

As Nigeria experienced changes in government, the goals of education were deferred but not cancelled. In 1977 the "Nigerian Education Report" was published based on the recommendations from "a seminar of experts from a wide range of interest groups within Nigeria [that] was convened" in 1973.[15] They recommended a curriculum change to the U.S. model of 6 years primary, 3 years junior secondary and 3 years senior secondary with a four year bachelor program. This change to the 6–3-3–4 model was fully realized by 1990. Initially it appeared that the beginnings of an indigenous education revision were in full swing. Textbooks were revised with cultural studies offered in the junior secondary levels and national history and government offered in the senior secondary curriculum. In the 1980s economic and political instability made an impact on Nigeria's education system and much of the mandated curriculum fell to the wayside. While it remained in the National curriculum, it was often not offered owing to lack of resources on the ground. The National curriculum further developed to create two clear tracks in Senior Secondary Education, the Arts and the Sciences. These tracks worked against the inclusion of indigenous knowledge in the curriculum.

Nsukka LGA maintains the national curriculum as established in 1977, but the indigenous education system that Azikwe envisioned is far from present. Students enroll in either "Arts" or "Science" curriculum tracks. Science is given the greater status, as it is designed to prepare for higher income occupations, such as medical doctor, law, accounting, engineering etc. These occupations were the most often sought after by the student participants. The senior secondary curriculum is almost stripped of social studies content; students reading "Science" take courses in the Sciences, Math, Agriculture, English, and Christian Religious Knowledge. These subjects are taught from Western knowledge bases. There is little to no opportunity to introduce Nigerian or African history, geography, cultural studies etc. It is not surprising that the Arts students are the minority in the secondary schools. In terms of core courses required by all senior secondary students, there was limited opportunity to include indigenous social science, much less indigenous sciences. According to a recent national curriculum guide all senior secondary students were required to complete "One [course in] Biology, chemistry, Physics or Integrated Science" and "One [course] in Literature—English, History, Geography, or Social Studies."[16] Science courses were consistently offered at both schools, but history courses were not always offered. History coursework included Nigerian and African history that covered ancient history through independence.[17] During this study the researcher was not able to collect or buy a history book at either school, but was given the titles of the ones they normally used. Alor Uno faculty reported that teaching history had not occurred in the senior secondary school in recent years. History was offered to art students at Shalom Academy, but again it was taken by a minority of students. Geography included the earth science aspects in the first year and human or cultural geography in the second. It was reported that many students did not take the second-year course. These courses fell under the Arts track, so again a minority of students took them. In contrast all Science track students' coursework included Biology, Physics, Chemistry, and Agriculture Sciences.[18] Science and Arts tracks were required to take Christian Knowledge at both schools. Within this context, the influences of Western science and Christian curriculum interacted to create a worldview that did not actively acknowledge or promote the study of indigenous science. Participants' perspectives on these influences created a continuum that indicated majority acknowledgement and acceptance of Western science and Christianity and minority acknowledgement or acceptance of indigenous science and traditional religion.

PARTICIPANT PERSPECTIVES

Many student participants commented on the issue of traditional religion and its related elements, in particular traditional health care, in the process of analyzing the effect of formal education on indigenous culture norms and values. Responses from two questions used in the student protocols demonstrated a shift away from participation in an African medical model. Much of this shift was directly tied to the participants' identification as Christian. Student participant 14S–F's comment is representative: "My grandfather again he involved himself in all these ritual things in the village and he was one of the native doctors in our village and he expected my dad to become one, but my dad refused. . . . My parents, they are Christians." Student 29A–M's comment was, "now we are Christian, we are no more doing that. That's their belief that the old people believe, but now we don't believe in that thing again. Now our parents have learned to be using all these medicines, but in the old they use leaves . . . Herbs and leaves to cure person, but now we goes to hospital. We are no more like them." All student participants shared these sentiments. Only one student indicated that his parent was a traditional believer, but interestingly indicated that his father recently converted to Christianity due to illness. He stated, "my grandparents they worship idols and my dad because of the title that his own dad got for him, so he now continues in that worship and all of that. When times go they now continue to convert, even though my dad is sick now, but he is still believing in God for total healing. So he has promised that when he recover that he will start going to church. But I don't see anything else, because all the titles that the other man wants for him, he is still holding it." Even though this student indicated some doubt that his father would totally convert to Christianity due to other traditional titles that he continued to hold in the community, it was clear that he had turned toward "Christian" prayer as an avenue of healing, which indicated a shift away from reliance on a dibia for health care. Whereas the available statistics indicated that at least 28% of the community population identified as traditional believers, all student participants identified as Christian.

In interviews with parents the shift from traditional health care to a Western model also aligned with identification as Christian. Parent participants validated their students' statements that they were Christian. Their interview responses also indicated the shift away from contact with a dibia or herbalist based on their self-identification as Christian. EE's response, "Traditional doctors I feel everything he does is from is the devil's work," was representative of parent and adult participants. Many parents/guardians/adults indicated valuing Western as "modern" with indigenous as supplemental to the "modern," as Stella Nwadinobi explained:

> Yes you know that time now when we were young we lived in the village so they value this local herbs much, some of them they prepare them by themselves and it works but now they are so much inclined with this modern treatments. My mother-in-law was a nurse and because of that once even if I have a headache now I will rush to the hospital to see a doctor so now I prefer seeing doctors than, I don't think I have anything to do with local herbs now.
>
> E: But if given a choice will you prefer to merge the two like using the local herbs where you know that they are efficient and then visit the doctor when you feel that one is better?
>
> I think to merge the two is better because there are some local herbs that work very well, so if we merge the two we will now know the one that is more efficient than the other or if we merge the two because some doctors . . . doctors they use both the local ones and the modern ones and because of that people rush to their clinics so left for me I will prefer using the two.[19]

These comments corresponded to the student participants' comments on movement toward acceptance of a Western medical model over an indigenous medical model based on their identification as Christian. This affirmed the notion of Christianity as one origin of influence in this transition away from an African medical model.

CURRICULUM RECONSIDERED

An analysis of the schools' curriculums provided insight as to why we could not find traditional believers or students who indicated awareness or involvement in traditional practices beyond participating in traditional weddings or yam festivals. Two major components of both schools' curriculums influenced students to endorse Christianity and Western science (including medical knowledge). The endorsement of Christianity provided

Figure 11.1 Shalom Academy, Nsukka, Nigeria (September 22, 2004). Photographer: Jamaine Abidogun.

the alternative belief system to endorse Western science, which includes the Western medical model.

The historical context effectively demonstrated this marriage between Christianity and Western knowledge as the two were introduced simultaneously in the form of formal Western education to the Nsukka area. As noted by A. E. Afigbo, "Formal education was first introduced in Nsukka area in 1912 when a school was opened at Uvuru in Uzo Uwani Local Government Area [later Nsukaa LGA] by the Catholic Mission." He further documented remarks made in 1958 by a first generation Western educated Nsukka leader, Chief D. C. Ugwu as follows: "Ugwu saw the advance of education through the agency of the Christian Missions as revolutionary because within less than half a century the Roman Catholic Mission and the Church Missionary Society [Anglican], had, between them, taken the light of education and Christianity to every town and village in Nsukka."[20] From the beginning the residents of the Nsukka area viewed Christianity and Western knowledge as part and parcel of the same phenomenon—formal Western education.

This integration of mission work and education in the Nsukka area is also demonstrated by colonial government documents. One such document addressed to the Eastern Provinces on March 27, 1942 stated, "The Board decided to recommend to Government that a period for Religious Instruction should be included in the normal time-table of all Government and Native Administration schools, and that there should be freedom of entry to all such schools for duly accredited representatives of approved missionary societies [excluded Pentecostal]."[21] The Eastern Provinces answered in the affirmative with reservations as they stated, "With reference to your endorsement . . . I have to inform you that while the Awka native Authorities expressed no objection to the proposal, the Enugu and Ndienyi Councils were at first averse to the proposal on the grounds that religious instruction is already provided by teachers. . . . They did not wish to press their objection, however, and finally agreed to the proposals though with some misgivings."[22] Their misgivings had nothing to do with opposing religious study in a government school, but with who would teach Christian religious study, i.e., that a denomination other than one that their teachers currently represented might gain influence over their students.

It is not surprising within this historical context that post-colonial Nsukka has not strayed from the integration of church and education. In both schools the order by the colonial government remains in effect, as two courses, Christian Religious Knowledge and Moral Instruction, are required curriculum for all students. There are no alternate religious studies courses in either school's curriculum. The connection to an endorsement of Western knowledge is therefore without question, as Science track students must take a comprehensive list of Western based science courses along with their religious instruction as listed above. Several student responses demonstrated an endorsement of Western medical science, such as 21S's statement, "To be a medical doctor . . . It helps me in some aspects, like for instance science, that is the main course." Or 25S's statement that, "I want to achieve my dream of being a medical doctor. . . . Science subjects I do. It help[s] me to know a lot about being a medical doctor. I want to be a medical doctor." And 29A's comment, " My secondary education help me like if anything I want to be in future it will be from this secondary education that I will know the idea of all that thing. Like if I want to be a medical doctor or an engineer from this place I will lay the foundation." Overall comments from student participants indicated that Western science knowledge was the "proper knowledge" and was indispensable to their future. By default this perception marginalized indigenous African knowledge and at best placed it as second to Western science in value.

The shift away from an indigenous African medical model began in the early 1910s, so it was not surprising that it was reflected in adult participants' acceptance of Western science as superior to indigenous science as it related to the medical field. Adult participants' perceptions were also influenced by similar curriculum, as they all indicated completion of at least a secondary education. Many parent and adult participants also demonstrated an endorsement of a Western medical model based on the influence of Western science knowledge, as JU commented: "So we only go to the chemist when we want to buy drugs and not for treatments or prescriptions . . . traditional medicines does not have dosage." UP expanded on statement, commenting, "You know nowadays, now that we are educated we know that a medical doctor knows better about health, he can go to laboratory and find out what is wrong with you before administering a drug to you." And EG explained, "I prefer Western medicines because . . . not knowing what it is . . . if you take the medicine without knowing the cause of the headache . . . you might get started with the wrong treatment." Just as the students marginalized the value of an indigenous African medical model, so did parent and adult participants. These comments by parent and adult participants also validated the second origin of influence as Western science curriculum that created a shift toward endorsement of a Western medical model.

In some instances the endorsement of Western medicine could be connected to changes in indigenous environment, i.e., changes in diet and exposure to new chemicals and other potential biohazards. Whereas it is certain such changes could coincide with increased use of Western medicine, this reasoning was not evident for study participants based on their behavior and responses. Through observation and interaction it was established that most families in the area typically prepared and ate indigenous foods as their regular diet. Also, participants' routinely identified traditional ailments and afflictions in the interview responses, i.e., malaria, fever, stomach upset etc. These observations and responses ruled out the alternative possibility that Traditional medicine was not suited to meet current needs.

The indoctrination process to accept a Western medical model as demonstrated by participant responses was almost total. Whereas Christian doctrine and a Western science curriculum taught through formal Western education were the dominant influences within this community, there were some exceptions to the perceptions held by these participants in the wider formally trained community. One such exception was an effort by an international nongovernment organization, Bioresources Development and Conservation Programme (BDCP) and its partner company, International Center for Ethnomedicine and Drug Discovery (InterCEDD), both officially incorporated in 1999. These organizations had central offices in the United States with laboratory facilities located in Nsukka. Their common goal was to secularize indigenous African medical knowledge so that it could be understood and used by anyone regardless of their religious affiliation. In an interview with A. Ozioko, the head researcher at the Nsukka location, he demonstrated the influence of his Christian upbringing as identical to that expressed by other participants. He explained: "When I was young, I am a born Christian; so my mother and my father are all Christian and then as such I grew up in such belief that we don't visit these people [dibias/herbalists] because due to their practice there. Because their practice in that doing things is so suspect to . . . our religion due to the way they practice it. You see some of them they say you produce hen, goat, they kill it there, spread the blood; they draw blood; which our people [Christian] was against.

Figure 11.2 A. Ozioko, biochemist, and W. Nwaefido, research assistant, at BDCP/InterCEDD laboratory in Nsukka, Nigeria (May 2005). Photographer: Jamaine Abidogun.

That's our religion. So in my youth I don't visit any of the herbalists."[23] His father's unique experience in his formal education provided an avenue for this researcher to value indigenous African science equally with Western science, as he further explained: "my dad had some knowledge of these natural drugs, which was if we develop some fever, he could go to the bush, collect some drugs and prepare and then administer to the children, if they have some fever or convulsion . . . because then my father was very close to a teacher who was a herbalist, but his own practice was a very decent [Christian] practice. They never go for this traditional [sacrifice of] cow or goat."[24] Whereas it is possible that he might have pursued this course of research without such an experience, it was this experience that he credited with providing him the necessary insight to pursue his career.

A majority of parents and adults indicated that they utilized common indigenous medicines in their daily lives even though they did not visit dibias. For example, one parent and professor at the University of Nigeria, Nsukka, related that her husband had stomach problems and visited the medical doctor on more than one occasion. None of the medicines he provided helped with her husband's condition. They finally did seek the advice of an herbalist in their home village and were given a particular bark to boil in water to drink as a tea. There was some concern about the question of dosage, as the herbalist instructor said, "drink one tumbler (glass) full," but failed to say what size tumbler. They started the dosage with the smallest tumbler they had and found a dosage that worked for her husband. As educated individuals they utilized indigenous knowledge to fill in the gaps in available Western medical practice. Neither of them knew the scientific (Western) name for the bark, but they did report that it was effective.[25] Their approach to blending indigenous knowledge and Western education was to remain active in the traditional councils from their home villages so that they could benefit the wider community. This benefit was the result of their ability to accurately represent accepted cultural practices, including common treatments for ailments, while providing information from their Western educated backgrounds that might compliment or provide information that when synthesized with indigenous knowledge could help others.

Other parents and adults, while not as purposeful in their approach to learning and utilizing indigenous knowledge along with Western knowledge, did indicate recognition and use of indigenous knowledge as a necessary part of their lives. An interview with B Agwuja provided a full explanation of how many members of the Nsukka community blended medical sources. The transcript follows:

> *2. Do you go to the hospital for health care or do you see the traditional doctors when you are ill?*
>
> It depends on the kind of sickness but we do go to hospital but if it is something that needs traditional care like poison or something like that you will go to the native one for that.
>
> *3. How do you decide when you need traditional medicine or when you need to go to the Hospital?*
>
> That is what I am saying that it depends on the kind of sickness the person have, . . . as you know this time development is everywhere but

sometimes they do several tests on someone and at the end the results can not be carried out and the person is still feeling sick, so you go to this hospital, no result and to this hospital, no result. So they will advise the person to go back to the village and use the traditional medicine and treat it and see if it will get better then the person if he feels better should treat it with the traditional medicine and is not supposed to go back to the hospital since he has visited many hospitals and yet no progress.

3a. Are the herbalists and dibias the same?

Yes, [the medicines are the same]. You know like in the market were they sell traditional medicines like in the bottle that one can just buy, are they the same medicines that one will find in the dibias' place?

Yes it is the same thing. Like now development reach everywhere in the olden days they do not sell traditional medicines in the market but now they do not have enough people that come and request for these things and you know they feed on them [so] they decided to bring them to the market and sell for those that are in need of it . . .

3b. When did this start happening, that they started bringing it to the market?

It has been long now you know in the olden days, people used to sick but now development started, let me say it is going more than 7 to 8 years that people started selling in the market and spreading it everywhere, telling people this is the work of this one and the work of that one. And this time again some children of God they do not want to go to the herbalist [dibia] because when you go they normally tell you that you should bring one fowl, bring kola and so, so and so and because they do not want to go against their faith by going there they will remain. It does not mean that they are not feeling sick but they do not want to go to them.

3c. They just want the medicine?

Yes, those church people that do not want to go to the herbalist to request for something if they see it in the market like that they can buy it with their money and they will feel relaxed that they did not go to the dibia's place.[26]

These examples demonstrated the possibility that others, if properly educated, could also benefit from the separation of religion and scientific knowledge bases. In fact much of BDCP's and InterCEDD's operations are geared toward that end. They meet with healers (including dibias) across Nigeria to collect "natural drugs," analyze their active ingredients, assess effective dosages and help healers to learn how to process and market these natural drugs effectively in the local market. These goals were apparently beginning to be realized in Nsukka area, as a few of the parent and adult participants attested to the shift from traditional medicines being only available through the dibia to becoming available in the local market by herbalists and in a few local clinics. Ms. Agwuja's interview and Dr. Okoye's experience and perspective are representative of the need to have access to indigenous knowledge, not to the exclusion of Western knowledge, but to benefit the most from both knowledge bases. Also the influence of indigenous knowledge was not completely prohibited from influencing future generations, as ST summed up the benefit of combining African and Western science, stating, "I think to merge the two is better because there are some local herbs that work very well, so if we merge the two we will now know the one that is more efficient than the other or if we merge the two because some doctors . . . they use both the local ones and the modern ones and because of that people rush to their clinics so left for me I will prefer using the two." It was encouraging that the older generation was willing to identify the value of indigenous science although with conditions that Christianity was upheld and Western knowledge incorporated.

CONCLUSION

The caution remained with the current generation of secondary students, as 28S so pointedly stated: "They [parents] will not contradict with me because they know that I read about that"; or 17A's statement, "Yes, we come to learn . . . , but you can't learn in the home; nobody can teach in the home." This perception established the authority of formal Western education and therefore Christian knowledge and Western science knowledge as superior to all other types of knowing. This question in the research protocols was designed to identify the integration of Western academic knowledge with indigenous knowledge, but the finding was dismal. In 57 responses, participants indicated that at home if they presented knowledge that was learned in school, it was accepted above the knowledge or practices of their parents without question unless the parents' knowledge had been acquired from formal education.[27] It was this warning sign that indigenous knowledge, historically taught orally by age grade, was being marginalized and eventually lost that motivated the writing of this chapter. We can learn from efforts such as the BDCP's and translate such examples into a revision of secondary curriculum to acknowledge diversity in religious belief systems beyond the middle school social studies "official statements" that Christianity, traditional and Islam religions coexist in Nigeria. Then the curriculum could include indigenous sources of scientific knowledge such as African medical models. Second, Western educated participants must reclaim and possibly relearn indigenous African knowledge. In this case study, Igbo Nsukka indigenous knowledge has survived almost entirely by the efforts of non-literate, traditional believers of this community. The BDCP supplied a promising example of how the people of Nsukka may add through reform in the school curriculum to this effort of preservation and integration of traditional medical knowledge.

In conclusion, Christianity and the formal Western science curriculum supported a shift away from an indigenous African medical model. To this end, formal Western education was an effective tool in indoctrinating students to believe Western science (medicine) and Christian religion were superior to African science and traditional religion. In a world were globalization is the trend it is vital to take the time to review and assess the trade-offs between a modern but often Westernized knowledge base as provided through formal Western education models and the maintenance of the cultural integrity of the society. Without such consistent review and assessment of issues such as these, the often overbearing emphasis on modernization in its Westernized form will continue to pervert and slant the cultural fabric of African societies.

NOTES

1. (Okolo, protocol B, 2004)—a female student at Shalom Secondary School, Nsukka, Nigeria.
2. Dibia refers to an Igbo priest or priestess who more often than not also possesses local medicinal knowledge for physical and psychological ailments. An herbalist in the Igbo context does not have the spiritual knowledge or position of a dibia, but does have medicinal knowledge for physical and sometimes psychological ailments. See Francis Arinze, *Sacrifice in Igbo Religion* (Ibadan, Nigeria: Ibadan University Press, 1970); Edward Ilogu, *Igbo Life and Thought* (Onitsha, Nigeria: University Publishing Co., 1985).
3. Maduka S. Onyishi, *Education and Community Development: A Case Study of Nsukka town, Nsukka L.G.A. of Enugu State* (BA Special Project, Dept. of Sociology and Anthropology, University of Nigeria, Nsukka, 1999), 56.
4. Special thanks and appreciation is offered to the following people: Dr Egodi Uchendu who collaborated on the site coordination for student interviews and conducted some of the adult participant interviews; Dr. Victoria Onu and Mrs. F. I. Agu for allowing me access to their schools, Shalom Academy and Alor Uno Secondary School, respectively; and to Ms. Winifred Nwaefido who assisted in the student interviews with translation and transcription.
5. Protocols are included in the index. All participants are identified with pseudonyms to protect their identities and privacy.
6. Critical theory, also radical critical theory as described in Henry Giroux, *Theory and Resistance in Education* (London: Bergin & Garvey, 2001).
7. Renee Shilling, "African Women and Spirituality: Connections between Thought and Education," in *Expanding the Boundaries of Transformative Learning,* ed. E. O'Sullivan, Amish Morrell, and Mary Ann O'Connor (New York: Palgrave, 2002), 152.
8. George Dei, "Spiritual Knowing and Transformative Learning," in *Expanding the Boundaries of Transformative Learning,* ed. E. O'Sullivan, Amish Morrell, and Mary Ann O'Connor (New York: Palgrave, 2002), 123.

9. J. E. Adetoro, *The Handbook of Education in Nigeria* (Lagos, Nigeria: Town & Gown, 1966), 4.
10. Wolf Roder, "Magic, Medicine, and Metaphysics in Nigeria," in *Sacred Realms: Readings in the Anthropology of Religion,* 2nd ed., ed. Richard Warms, J. Garber, and R. J. McGee (Oxford: Oxford University Press, 2009), 289.
11. Arnold Guy, *Modern Nigeria* (London: Longman, 1978), 104.
12. Institute of African Studies, "University of Ibadan and University of Pennsylvania Summer School Program," (Ibadan, Nigeria: University of Ibadan Press, 1989), 2.
13. Nnamdi Azikwe, *Renascent Africa* (London: Case, 1968), 144.
14. Michael B. Adeyemi, "Nigerian Social Studies," *The Social Studies* 77, no. 5 (1986): 202.
15. Federal Republic of Nigeria, *National Policy on Education,* 3rd ed. (Yaba Lagos, Nigeria: NDERC Press, 1998), 5.
16. Ibid., 20.
17. L. M. Okafor, *History for Senior Secondary Schools: Nigeria*, Books 1 & 2, 2nd ed. (Onitsha, Nigeria: Jet Publishers, 2004); L. M. Okafor, L. E. Amadi, and L. O. I. Ogueze, *History for Senior Secondary Schools*, Book 3 (Onitsha, Nigeria: Jet Publishers, 1990) were history textbooks available in the Nsukka LGA markets.
18. Alor Uno Senior Secondary School posted curriculum for Science students (2004–2005).
19. Interview Stella Nwadinobi, parent of Nsukka students, conducted by Drs. Egodi Uchendu and Jamaine Abidogun at Nsukka, Nigeria (March 20, 2005).
20. Adiele E. Afigbo, "Nsukka Zone: Culture, History and Development," in *Nsukka: Development and Challenges,* ed. Damien Opata, Julius Ozioko, and Chuma Eze (Nsukka, Nigeria: Development Challenges; Nsukka: Adada Investment Ltd., 1997), 143, 150.
21. Education Department, Lagos Nigeria, "Religious Instruction in Native Administration and Government Schools," No. D.E. 354/4, Enugu State Archive (March 27, 1942).
22. The Divisional Office, Awka, Nigeria, "Religious instruction in native Administration and Government Schools," No. Aw: 124/571, Enugu State Archive (May 14, 1942).
23. Interview with Alfred Ozioko at the InterCEDD office in Nsukka, Nigeria (May 2005).
24. Ibid.
25. Discussion with Dr. Uzo Okoye and Dr. Jamaine Abidogun at Nsukka, Nigeria. Their children attended the University primary and secondary schools (2005).
26. Interview with Beatrice Ugwuja, parent of Nsukka students, conducted by Drs. Egodi Uchendu and Jamaine Abidogun at Nsukka, Nigeria (March 16, 2005).
27. Summary analysis of all responses to Protocol B, Question 6. Is there anything that you do or learn at school that does not agree with or match what you do or learn at home? Please explain.

12 Depictions of Sociopsychological Dimensions of the Niger-Delta Crises in Recent Nigerian Drama

Alexander Kure

INTRODUCTION

This chapter considers the adroitness of literature, using selected recent dramatic texts in Nigeria, to interrogate the sociopsychological dimensions not often explored adeptly by other domains of scholarship with a view to unravel the unknown motives that have persistently ensured an inadequate interpretation, hence non-resolution, of the Niger-Delta crises in spite of concerted efforts by government, the people of the Niger Delta and other spirited individuals and organizations. The chapter posits that discourses so far on the causes and nature of the crises in the Niger Delta have remained inadequate inasmuch as they dwell on the superficial or literal aspects of the dimensions of the crises at the expense of the more expressive motives and dimensions: the sociopsychological dimensions. To this end, the chapter is foregrounded on the premise that until the sociopsychological dimensions are explored, interpreted and understood in addition to the other dimensions being discussed, the discourses in respect of the crises in the Niger Delta will continue to remain inadequate.

In light of the focus expressed above, the chapter is divided into four subsections. The introduction combines the review of some of the literatures of discourses by various scholarships on aspects of the study in question as well as a conceptual framework. Further, an in-depth exploration of the plays (basically contextual analysis at the literal and symbolic levels) is done with a view to conceptualize and prove the basic assumption of the chapter. The selected dramatic texts are J. P. Clark-Bekederemo's plays *The Wives' Revolt* (1991)[1] and *All For Oil* (2000)[2] and Ahmed Yerima's award winning play *Hard Ground* (2006).[3] For the avoidance of any doubt, the assumption is that besides other domains of scholarship, it may be different genres of literature that can be used to appropriately interrogate and understand the causes of the basic issues that have defied efforts at resolving the lingering violent crises in the Niger Delta.

The assertion of the chapter presupposes that interpreting the changing and resilient nature of crises in the Niger Delta calls for a totally new approach. It argues that a narrow focus on material resources had prevented analysts from identifying the real sources of vulnerability or resilience in a population. Therefore, the chapter uses a conceptual approach that pays attention to the sociopsychological bases of stability in the Niger Delta. It addresses why crises remain pervasive in a particular community going through the same problems like another community that had its crises resolved using the same models. The chapter conjectures that such failures may be accounted for by the existence of complex situations that are best explained only by a composite model of human security.[4] By the assertion above, it is assumed that for crises in a community to be resilient, the society needs not necessarily to be rich. Instead, what it needs is a minimum level of survival. If no such level exists, development gains will be short-lived and illusory and the potential for conflict will be high. By contrast, if minimal material inputs can be guaranteed and if efforts can be made to enhance basic social coping capacities, communities will be more stable and less prone to fragmentation, violence, and any form of atrocity. In particular, it is identified that psychologically, if three areas of stability are identified—home, community and the future—instability will not persist. When these attachments are undone, as they are when populations are uprooted and dislocated, when families and communities are broken up and arbitrary violence and discrimination render the future distant and unpredictable, individuals then turn to other sources for participation, recognition and empowerment. These sources take the form of identity groups formed around ethnic groups, religion, geography or age and are characterized by an ideology of resentment and aggressiveness against established institutions and processes. When these become manifest in sections of the population, credible dispute resolution becomes unlikely and the paths to violence and disorder become manifold.

The conceptual framework for this chapter proposes a new concept of interpretation and shows how it can be applied to explore the persisting crises in the Niger Delta. Hence, focus is more on the individual rather than the general community and pays attention to the sociopsychological areas of resilience (home, community and the future). In context, therefore, the chapter shows, through its explorations of the selected plays, that discussions on the measurement strategies take cognizance of dynamic changes in the home, community and the future to show the indeterminateness of the crises. These will assist the chapter to capture the levels of detachment or dislocation from home, the extent of destructive tensions between community groups, and the degree of pessimism about the future. In espousing this view, this discourse is supported by the expressive perception of Marris when he says

> Our theories of human behavior split into largely independent systems of thought, psychology and social science. . . . We rarely explore the interaction between each unique human actor and social systems of which she or he is part. Yes, surely this interaction ought to be at the

> foundation of any theory [of] human behavior. How can we begin to understand ourselves except as a network of patterns of relationships which each of us is constantly engaged in creating, reproducing, and changing? We need a way of thinking about the interaction between unique human beings and the social relationships they form, not only our theories are crippled without it, but because without it we cannot articulate clearly many of the gravest causes of social distress.[5]

The Niger-Delta is the world's largest wetland (22,600 square kilometers). It accounts for 7.5% of the land mass, and covers a coastline of 320 kilometers, about two-thirds of the coastline of Nigeria. It may be the largest mangrove forest in the whole world (12,000 square kilometers). Ikpatt and Scott[6] say the Niger-Delta comprises of six or seven states, but in line with Ibeanu,[7] Opara-Nnadi[8] and Nkoro,[9] one can submit that there are nine states: Abia, Akwa-Ibom, Bayelsa, Cross-River, Delta, Edo, Ogun, Ondo and Rivers. The estimated population of the area based on Nigeria's census figures for 2006 is about twenty million, consisting of over forty different ethnic groups with three hundred communities that speak two hundred and fifty different languages. The predominant occupations of the people in precolonial and colonial times were fishing and farming. Since the successful exploration and exploitation of oil in commercial quantity starting at Oloibiri in present day Bayelsa State by the Anglo-Dutch group Shell D'Archy in 1956, the wealth of Nigeria as a nation has principally been in the Niger Delta. Without the Niger Delta, Opara-Nnadi says "Nigeria will be gasping for breath to survive."[10] This is so because the Niger Delta accounts for about 13% of Nigeria's gross domestic product and about 90% of government revenues annually; hence, it remains central in providing the funds to ensure the functioning and activities of government in Nigeria.

However, in spite of the role played by the Niger Delta to sustain Nigeria's economy, the area remains perpetually exploited and virtually undeveloped. What accounts for this state of affairs is summarized by Utomi to include six cumulative key factors that have historical antecedents: "the challenge of minority rights in a multi-ethnic country; a history of poor or bad governance; the allure of crude oil stealing; the challenge of principles-based nation-building with regard to fiscal federalism; the 'criminalization' of politics and the enslaved nature of the economy which limits the trickle-down of benefits to the community."[11] Therefore, the combination of the above and many more has rendered the area to be the most undeveloped and hazard prone in Nigeria. To be specific, there has been very little substantive development in terms of the provision of water, social services, poor roads, hospitals, education, lack of viable employment opportunities especially for youths and environmental degradation with its resultant health and economic hazards. In a nutshell, the Niger Delta suffers from an acute absence of aggregate or quantifiable economic growth and development of whatever nature.

It is pertinent to elaborate on the issue of environmental hazards caused by the activities of oil companies in the Niger Delta by paraphrasing Ibeanu to situate the discourse in proper perspective. He says, among many other things, that "prior to the Bonny project, the practice of oil companies was to flair the gas." This meant "enormous amounts of natural gas were flared annually, usually in the vicinity of human dwellings." For example, he notes that "in 1982, Shell, Gulf, Mobil, Agip, Texaco, Pan Ocean, Ashland, Phillips, Tenneco, and Elf flared over thirteen million cubic meters of gas in 145 communities in Nigeria." Specifically, "Shell" is singled out as the "worst culprit," having "flared nearly seven billion cubic meters of gas in that year alone. In one case, in the town of Sapele, Shell burnt over 944 million cubic meters of gas." Further, he confirms that "in 1991, Nigeria exceeded the world average for natural gas flared by seventy-two percent. In that year, while the world average for gas flared as a percentage of total production was four percent, Nigeria flared seventy-six percent of the total production." The cumulative effect of the activities above "apart from the huge quantities of greenhouse gases that gas-flaring pumps into the atmosphere," is also that "constant flares affect both wildlife and human beings negatively" because "at temperatures of up to 1,400°C, the 'sauna bath' effects that gas flaring produce make living in many communities nearly impossible."[12]

The consequences have become too grave to bear by the people because of the inability of successive governments, starting from the colonial to date, to resolve the abysmal condition of the people owing to the multiplicity of factors that are prevalent in the face of the continuous exploration and exploitation of oil in the oil-rich Niger Delta. In more specific terms, Osadolor asserts this gave birth to the following reaction:

> [T]he crux of the issue that is of interest to us is whether or not the context of the understanding of the problems in the Niger-Delta derives from the debates on the National Question . . . the grievances, demands and struggles in the Niger-Delta which derive from the deprivations of their resources and opportunities for development is directly linked to the National Question.[13]

The issues have reached the dimensions of being tabled as a National Question, which makes one wonder at the real state of affairs in the Niger Delta. The reason is not farfetched. The consequences of the continuous exploration and exploitation of the abundant natural resources coupled with the neglect of the area especially at the national level are the reasons for this. Meanwhile, discussion in this chapter does not lose focus on the fact that long before the advent of crude oil exploration and exploitation in 1956, infrastructural and many other problems had existed in the region basically because of the peculiar geographical location of the area. The area is swampy, with innumerable creeks and rivers. Transportation through this ecosystem, which is usually via rivers and creeks that snake through the dense, mosquito-infested swamps, is very difficult. Resultantly, there is a high incidence of malaria and other water-borne diseases in the area. Indeed, because the Niger Delta has remained a very sensitive ecosystem, the peculiarity in itself has been a major problem for developing the area. In addition, the seats of initial successive governments were either at Enugu, Ibadan or Lagos. The failure to cite such in locations close to the people in the Niger Delta to engender rural development until the successive creation of states and Local Government Councils contributed to distancing the people from development. The desire to earn attention did not only remain a mirage to the people but ensured the communities remained grossly undeveloped.

In spite of the existence of long-standing grievances, however, the exploitation of crude oil and the untamed degradation of the environment caused by deforestation and oil spillage have further fuelled the tenor of the crises in the region. The Niger Delta is being destroyed in many ways by the oil companies. Ibeanu says, "very high-pressure pipelines criss-across villages carrying crude oil, premium motor spirits, diesel, and gas. Spill and leaks from these pipelines and other installations that destroy wild-life, farmlands, forests, aquifers, and human lives." He notes also that "within the last few years separate fires have claimed hundreds of lives in the states of Delta and Abia." Additionally, the "discharge of refinery effluents that contain excessive amounts of very toxic materials like mercury and chromium into freshwater sources devastates the environment and threatens human lives." Finally, "badly constructed canals and causeways built to facilitate activities of oil companies have equally affected the hydrology of the region, causing flooding in some areas and water deficiency in others."[14] Pointedly, large forests have atrophied as a result. In addition, most of these artificial canals also let saline waters of the Atlantic Ocean into freshwater sources, thereby increasing the scarcity of drinking water and killing many species of plants, animals and fish.

The fact that there has been an apparent failure and insensitivity on the part of parties concerned to employ existing technologies to combat some of these ecological problems have led not only to turning the problems of the Niger Delta into a national question, which presumes a quest for dialogue, but to more violent agitations. Therefore, it is not surprising to hear of violent conflicts, sabotage of oil wells, vandalized pipelines, the kidnapping of various kinds, using unheard of methods, of personnel that work with multinational oil companies and even of the civilian polity to make the area inhabitable for oil explorers and exploiters. Indeed, some of the kidnappings have the trappings of political revenge, criminality and shear banditry. The failure of government to actually find a suitable way to combat the mounting pressure being put by the people of the Niger Delta to resolve their perennial problems other than attempting to resolve the crises by mounting militarization in the area, attempting to enforce justice and peace at the same time and the misguided use of the term "granting of amnesty," have further exacerbated the crises in the Niger Delta. Often times, government resolved to use its military might to quell violent situations in the Niger Delta without recourse to resolving the very apparent contradictions of life in existence in the communities. Often this has not gone down well with the people in the communities in the region who are daily confronted with mounting pressure on the quest for survival.

The failure to be attended to by governments especially at the national level over time has again brought to the fore the argument that the situation remains unchanged because it affects an area occupied by minority tribes because the center of protest and political activism has shifted to the western states in the Niger Delta, particularly Bayelsa and Delta Sates where the Itsekiri, Urhobo, Ijaje and Ijaw communities predominate. However, the argument here is that because the region parades the largest number of minority groups in Nigeria, it becomes crucial for government to reconsider its position to ensure peace and tranquility are maintained. To press this point home, there are continuous calls for an increase in the percentage allotted to states in the Niger Delta using the derivation formula, strict adherence to the practice of fiscal federalism and the complete political restructuring of what is called the Nigerian State. In addition, several ethnic nationalities in the area have made declarations that amount to bills of rights to press home their demands for recognition and attention.

However, in spite of the seeming abysmal picture that has been presented on the state of affairs in the Niger Delta, it is pertinent to reiterate that some concerted efforts, as it were, have been made to combat the crises in the Niger Delta by successive governments, organized groups, individuals and even indigenes of the Niger Delta. A few examples will suffice here. During the colonial era in 1947, the efforts at self-determination by the Niger-Delta Congress caused the creation of the Rivers Province. In addition, the effective articulation of the problems of the Niger-Delta at the Sir Willink Commission led to the Niger-Delta Development Board's incorporation into the 1960 Nigerian Constitution. In the post-1956 years, cases abound where the people of the Niger Delta sued oil companies operating in the area and got compensations and an increase in the provision for infrastructural facilities. Further, at different times, the Federal Government created a few states from the initial Mid-Western Region and also carved out some more states from those few states. In addition, it also created Local Government Councils and traditional institutions in the Niger-Delta area to ensure the people got some measure of meaningful grassroots representation and development. Government has also created the Niger Delta Development Commission (NDDC) and of late a full-fledged Federal Ministry of the Niger Delta to handle the affairs of the Niger Delta in pursuance of achieving the objectives of development, peace and stability. Interestingly, the activities of bodies like the Movement for the Survival of the Ogoni People (MOSOP), the Ijo Egbesu Boys and the Niger-Delta Volunteer Force (NDVF) have succeeded to push the people's plight and struggle to the international domain of discourse, thereby earning daily support in many ways for the people and their struggle. Although there was a failed attempt to convene a summit to discuss the Niger-Delta problem in 2007 by President Yar'adua, the Federal Government has stated severally its commitment to ensuring the provision of amenities to satisfy the various requests by the Niger-Delta communities to assuage their feelings of marginalization and neglect. Efforts by governmental agencies and spirited persons are known to have started and supported the activities of The United Delta Energy Development and Security Strategy (UNDEDSS), the Job Creation and Empowerment Summit among many other summits, workshops, conferences and seminars that were organized to find ways to resolve the crises generated by the various demands in the Niger Delta. However, it is only important to add that the aforementioned efforts to resolve the contesting problems in the Niger Delta are like the proverbial throwing of a drop of water into a roaring ocean, as far as knowledge of the problems that are at issue in the Niger Delta are concerned. Simply qualified, such efforts at resolving the persistent crises are just grossly inadequate and most often exhibit a lack of sincerity of purpose on the part of especially the government and the people in the affected region.

Looking critically at the scenario painted above, it confirms the chapter's contention that there is still a vacuum that has been created by the missing link in the discourses on the Niger Delta. Although scholarship like the fields of psychology, geography, political science, business administration, and economics may explore the sociopsychological dimensions of the crises, the efforts have remained inadequate. It failed also because such discourses are not situated at really interrogating the deeper aspects of human nature: how and why it functions the way it does in the face of mounting crises that require an urgent and amicable resolution of the issues in contest in the Niger Delta. Some posers will elucidate the position of this contention. For example, there are communities in different countries of the world that have had oil explorations and exploitations and equally suffered similar effects like those in the Niger Delta. Yet, such communities today are living in apparent peace even if in the face of a few crises. How were their problems resolved? Why is it that the situation in the Niger Delta in Nigeria is defying all the time-tasted methods at conflict resolution? Further, one wonders why, for example, the defenders and fighters for the rights of the people will turn against themselves and even kill one another in the process or why those persons live an affluent lifestyle different from those they are fighting for live in poverty. Interestingly, one wonders why the Niger-Delta crises failed to dovetail into some religious imbroglio, as it is likely to have been in all conflicts in say Northern Nigeria even when one of the most feared groups is led by a professed Muslim, Alhaji Mujjahajid Asari Dokubo. Or why is it that the act of kidnapping virtually became an act of sheer criminality for financial gains? Perpetrators kidnap just about anybody and request for the usual ransom when those persons have no link whatsoever with the issues that affect the transnational oil companies concerned. It is also crucial to wonder why government would even sponsor the activities of such bandits. Questions and observations like those above make the discourse to submit that the answers are to be found in an in-depth, sociopsychological exploration of those issues. It is further argued that the different genres of Nigerian literature offer the materials that can be used to confirm this assertion.

IN THE EYES OF LITERATURE: BEYOND HISTORICIZING THE CRISES IN THE NIGER DELTA

There were some literatures on the Niger Delta long before being popularized by and formally referred to as literature on the Niger Delta by the veritable poet, Tanure Ojaide. Early writers like J. P. Clark-Bekederemo wrote about the Niger Delta to mainly showcase their cultural significance such as marriage system, superstition, occupation, geography and so on. Yet such writings did not lead to mainstream attention of the problems in the area or sharper perspectives until Ojaide and others wrote about the issues. Interestingly, because literature that is socially committed is people oriented, its discourse centers on and revolves around the societal events that are literarily an interface with history. The interplay and use of history by literalists; however, does not boil down to mere facsimile historical or literal documentations and interpretations. Handling such events in the hands of a writer that exercises poetic license ensures issues are effectively (re)interpreted and evaluated to draw deeper meanings that would apparently not have been glaring should only raw historical data be considered. To that effect, a multidisciplinary approach helps to situate the discourse and argues on why certain lines of action come to play during periods of crises like in the Niger Delta. This approach is quite different from the way other disciplines have sought to interpret the crises. A content exploration of the selected texts would further elucidate the point being argued here.

J. P. Clark-Bekederemo's humorous play *The Wives' Revolt* (1991), divided structurally into six parts, literarily dramatizes what women see as injustice, degradation, subjugation and oppression, the seed of which is sown by an oil company. The people of Erhuwaren are given a certain amount of money in the form of a token to appreciate the people on whose soil they are exploiting crude oil. The money is shared into three parts and given to three different groups: elders, men of designated age groups and women. An argument that borders on the age-old custom of non-recognition of women in decision making erupts. This argument metamorphoses into a quest for recognition by the women. Because of that, women are presented as being dissatisfied; hence, they protest the sharing formula. They would wish the money were shared into two parts only. Okoro, the town crier, reports this when he says, "However, our women folk, led by a few reckless one's . . . repudiating this fair and reasonable distribution of the money, demanding that it should have been divided equal parts . . . they have refused to perform their civic duties and responsibilities."[15]

Men turn down the demand by women because of the culture of not recognizing the rights of women to decide on matters like the above in the society. Consequently, for the very first time and uncharacteristic of rural women cast in other works by this author where his women are often bound by old customs and traditions of never protesting decisions arrived at by men, they decide to refuse to perform their civic duties and responsibilities as wives as a mark of protest and feminine solidarity. This places all domestic responsibilities squarely on the shoulders of men. Although the men are angered by such audacity, they are equally happily aware that women possess the many goats that abound in the community. Therefore, men chose to hit back at women by enacting a law banning the movements of such animals. The reasons for the law are that women are accused of using the power of witchcraft to assume the insidious forms of goats to terrorize citizens of the town and to ensure that the droppings of such goats do not litter the environment again. Hence, any goat or such other animal found moving about in the community faces death.

The new legislation aimed at decimating the resolve of the women triggers yet another reaction by them. There is an exodus of women to Iyara (a neighboring village) as a continuation of their revolt. This action further places the execution of all family duties customarily/traditionally performed by women in the hands of men. When men realize the enormity of the functions left for them, which they had thought they could handle coupled with the futility of their efforts, a process of reconciliation ensures and the women return. Okoro authenticates this position in the play when he says

> No award is too much for this singular set of women, who, in fearless opposition to an unjust law, staged a walk-out on all men, bringing life to a standstill, not even sparing their children. . . . It is enough to say that husbands will continue to share with their wives whatever fortune comes their way. The story is therefore finished, it is finished![16]

Yet their return is incomplete because each woman returns with a venereal infection they obtain through the mischief of Ighodayen the common prostitute once popular with men from Erhuwaren. On the women's return nothing can be more oppressive again like the action of Okoro where he beats his wife Koko mercilessly because she returns from Iyara with an infection without giving her the opportunity to state her part of the story. Meanwhile, the law that triggers off the protest is cancelled, a new approach to use the oil money is worked out and a team of medics are brought in from Warri to cure the women of their infections. Interestingly, the play uses only three characters to tell the story even though one could read, visualize and feel the presence of other characters not identified by the playwright.

Beyond the suggestiveness of the play being an attempt to contextualize the quest for the acceptance of feminist ideals in a community, and the non-appreciation of the value of women in the society, the play explores a multiplicity of meanings that include the inherent sociopsychological issues in the community in the Niger Delta. In addition to the adduced meanings mentioned earlier also are greed, marital immorality, oppression and the problems caused by the oil company to an erstwhile peaceful community basically through handing out physical cash other than physical development. But the chasm between the wives and their husbands emanates from oil money, and this gives birth to results that become socially and psychologically devastating to both parties and indeed the whole land as the fight to confirm who should take the lion's share rages. Without mincing words, this has continuously been one of the issues perpetuating the crises in the Niger Delta. It is a truism that there are individuals and groups that received such largesse from the oil companies but failed to declare such or share it out equitably; hence, different forms of reactions have been engendered. Yet unlike the present situation in the Niger Delta, the community in the play becomes very conscious of the futility of their struggle; they embrace the path of reconciliation, beaming light on the lack of physical development in the land. This is a very new consciousness that bears resemblance to how the situation should have been resolved in the Niger Delta today. Because the people have become conscious of the need for physical development and of the insensitivity of the oil companies and of their right, the host communities can no longer be bamboozled into fighting over what is certainly a periodic largesse.

The message in the play is a progressive message to the community and the oil companies, maybe because of very obvious data available, particularly Shell, in their foolishness and oppressive gimmicks. Okoro's comment quoted extensively here lends credence to this claim. He says

> Erhuwaren! Ehurwaren! Listen carefully, all of you! Open your ears wide so that you may hear well. The town has been brought together; all quarters, sections and shades of opinions, as from this moment, are all reconciled and re-united with one voice. The unfortunate incidents of the last few days have shown how an idea, meant for the common good, if not debated in public and adopted by consensus, can give rise to dissention, and create havoc in a community, right down to the family level, even to the nuclear unit of one man one wife. Whose eyes, whose ears among us have not witnessed the truth of the matter these last few days? To bring peace and harmony back to town, back to the home, it makes any heart sweet all over to be able to announce that with immediate effect, even as I am speaking, the obnoxious law banning all manner of goats from our city and anywhere near the precincts is hereby revoked and consigned to that place where all unhappy things are better left and forgotten. . . . As for the original matter of the oil

> company money that started all this fire, let it be known here and now that it was not such a big sum of money. Certainly, not so big it was going to change the condition of our lives permanently better. It has left the poor, poorer, and the rich, perhaps, little richer as our oil continues to flow to enrich other people across the country. But that is another story. Right now, let nobody pour petrol on fire that is already dying down in all households. Since it cannot even build us a road or clinic, it has been decided by consensus that we open a school building fund with what is left of it for the benefit of our children. In education lies equity; lies equality.[17]

In comparative terms, the above statement clearly indicates the sudden radical transformation in Clark-Bekederemo's focus on his thematic concerns as a writer. Before *The Wives' Revolt,* his dramatic works have often dwelled on showcasing, in its entire entity, the very rich culture of his people in the Niger Delta. There is no gainsaying that the above formed a greater part of his entire efforts while actively in the academy. Examples of such thematic concerns abound in plays like *The Masquerade, Ozidi, Song of a Goat, The raft, The boat* and so on. The simple conclusion is that it indicates a kind of rebirth, expression of regret of erstwhile misdirected efforts or perhaps a class suicide on the part of the veritable and world renowned playwright. This rebirth in general is further accentuated and confirmed in the play *All for Oil.*

All for Oil postulates the story of the creation of Nigeria by the British after the Berlin Conference of 1884. The British created Nigeria to exploit the palm oil and other commercial resources in Nigeria at the time shortly after the end of the slave trade. It is instructive to note that that included the Oil Rivers Protectorate in the southern part of Nigeria. Although it was the work of Lord Lugard to form the political entity called Nigeria, it was Sir Tubman Goldie that formed the economic enterprise Royal Niger Company that had the monopoly to exploit in commercial quantities palm oil. Noting this development, the lead character in the play, Bekederemo (full name Ambakederemo) complains thus:

> Do you see how the white man has piled our plate full with problems? He has summoned a whole crowd of people, big small, each speaking a different tongue, into a single hall, which he calls Nigeria. How are we going to hear each other speak? And above all the din, he hands us a song to sing at gun-point. We are in trouble, my brother.[18]

Another related problem is that in the days of Ambakederemo, oil price was determined by the British through price fixing, which denied the Southern Protectorate the opportunity to determine the control of their resources. In those days the British determined the grades of palm oil (grade one, grade two and grade three). At the time the play is set and as at today in the oil market, it is the same sweet or light Bonny and the heavy kind that determine oil prices too! This attitude angers Bekederemo; hence, he observes sadly, "Yes, a curious market they have established for themselves, buyer dictating to the seller the price of his goods."[19] It is definite that Bekederemo is not in the league of those agents that are being used in the British divide and rule system to decimate the natural resources of the people using local community agents. For example, the character Egbe complains against one of the collaborative sellouts, Dore, who is the Honorable Paramount Chief, and also the Olu of Itsekire, Urhobo, Ijaw and Warri thus:

> It's not just Ijaw and Urhobo lands here that he is signing away without the people's knowledge and consent. He has done it at Okere, at Sapele, and now he is signing away more for the new place called Alder's Town, all in the name of our Olu that every child knows died before his grandparents were born. In return, the white man carves for the man a portion of the land that can hardly take the quarters for his cook and steward.[20]

The above assertion on the role being played by Dore is confirmed by him when he says

> I'm sure they are Sobo—not from Agbassa, I hope I have taken new parcels of land from them without their knowledge and consent. The white man seems to need more land every day, not just for his stores, I see, but for offices, barracks and quarters for his supporting staff from Sierra Leone and Gold Coast.[21]

In the play, Dore is cast as the protagonist of the principal political agent for Lord Lugard. Historically, Dore Numa led the British in their wars with Nana Olumo in 1884 and Oba Ovonramwen in 1887. It is Bekederemo who is the preeminent trader in palm produce in the then western Niger Delta, from Warri to Patani into the present Bayelsa. He is the major middleman. He buys palm oil and kernel from the local producers—Isoko, Urhobo, Ijaw, Aboh—and sells to the Royal Niger Company, precursor of the United Africa Company. As a principal merchant like many others that littered the economic terrain of Nigeria at the time and as one that opposed any sense of unfair play in trade by the Europeans, he becomes the traditional nationalist that fights the likes of Chief Dore and it is from the ashes of whose patriotic struggle the likes of Azikiwe and Awolowo emerged.

The trajectory of the story in the play is that Nigeria was a market for the British when they gave up the heinous slave trade to trade in palm oil and now crude oil. Many years after, what is keeping Nigeria together is oil, the playing ground being the same imperial Nigerian state, the players, the multinationals now led by the Royal-Dutch Company, Shell for short, and at the bottom of the heap the poor people of the Niger Delta, left divided as usual and quarrelling among themselves. The play shows that it is the same old story with the same effects still being played out in the Niger Delta: the bottom line being the concerted efforts by individuals and groups of individuals to outsmart each other in the quest for patronage by the multinational companies and government against the cooperate interest of the people. The conflict being played out between Bekederemo and Dore Numa in *All for Oil* and men and women in *The Wives' Revolt* because of intrepid greed and carelessness are symptomatic of the situation that has been raging and that has made it impossible for the problems to be resolved. This position, as stated earlier in the essay, is a new pointer toward understanding the issue that is diametrically opposed to all other positions proffered by other scholarships.

The contentions that further expose the psychology of the people in the Niger Delta in the above historically based play are foregrounded on two premises. These become clear when brief historical analogies and their implications are critically examined. In a nutshell, the issues border on the emergence of what is called Nigeria today and discourses on its nationalistic struggles. Prior to 1914, there was no country called Nigeria. The inhabitants of the area along the historic Benin River and around the ancient Rivers Niger and Benue belonged to autonomous kingdoms, empires and emirates like the Benin Kingdom, Itsekiri Kingdom, Sokoto emirate, Oyo Empire, Opobo Kingdom and so on that were administered independently by their approved authorities. But for colonial intervention, these many pre-colonial setups would have metamorphosed into nations. It is crucial to state that European incursion for all intents and purposes was for purely economic reasons as manifested in the way and manner boundaries were drawn in Africa. This same pattern was used in all countries allocated to colonial nations after 1914. Because colonialists disregarded ethnic and cultural consanguinity in the process of parceling out areas, in the process, homogenous ethnic and cultural groups were split, thereby making kith and kin to end up in different nations. Examples of these abound at the border towns of Niger, Cameroon, Benin Republic and so on. Therefore, Nigeria on her part emerged from this situation as a mere accident of the same history.

At independence in 1960, the general proclamation seemed to have been one nation, one people and one destiny. Nigerians should have known that was a vain desire because it was clear that Nigeria was made up of different nations, several people with different destinies and destinations. The political compass used to create the first three regions—North, South and West—brought the first contradiction to the fore, hence the creation of the Mid-West region even when it was not located in the middle of the area west of Nigeria. However, the regions were potential nations, each with economic viability, cultural identity, ethnic homogeneity and to a large extent, similarities in the languages they spoke.

From the foregoing, first, the play states categorically why it is difficult for even J. P. Clark-Bekederemo, the playwright, to subscribe readily to the dominance of the Niger-Delta region by any other ethnic group in whatever name called and to even exploit the natural resources that belong to the people because as far as the region is concerned, it could have grown to become a very viable nation emerging from its traditional pre-colonial kingdom. Second, the play disabuses the historical notion often advertised that Nigeria's eventual independence was fought and obtained via the resisting efforts of founding fathers like Obafemi Awolowo, Sir Ahmadu Bello, Tafawa Balewa, Nnamdi Azikiwe and a host of other nationalists who hail from especially Nigeria's major ethnic groups. The play would prefer they are referred to as modern nationalists. The argument that supports this request predicates on the logic that bearing in mind the pre-colonial nature of Nigeria and the history of early European economic incursion, there is the need to trace the anti-European struggle against colonial economic incursions especially by understanding the mechanics of the trade that went on in economically rich localities. If the foregoing perspective is logical enough, then it would mean that the evidence that Europeans were resisted by certain pre-colonial communities, in the case of the example shown in the play, the efforts of Chief Bekederemo, means those communities and especially the persons so identified makes them to be qualified to be called nationalists too! Therefore, there is the need to straighten the records on nationalistic struggles in Nigeria to identify those traditional anti-colonial nationalists first. The failure at identifying such deep rooted psychological feelings as expressed in the play will continue to feather deep feelings of resentment at whatever efforts that are made to resolve the crises in the Niger Delta. Clark-Bekederemo is quoted by Nwosu in an interview granted at the instance of the Bekederemo Foundation and Museum in Kiagbodo to immortalize the founding father of the Bekederemo clan who stood against the shoddy business deals of the British and their corrupt middlemen thus:

> Who were the palm oil traders? Men like Bekederemo from Warri to Patani into Bayelsa. He was the big middleman in charge. He bought palm oil and palm kernel from the local producers from Isoko people, Ijaw people, and he sold to the Royal Niger Company which became UAC. That was how Nigeria was put together.[22]

His many other insightful comments on the neglect suffered by communities in the Niger Delta requests us to also further note the need to interpret the sociopsychology of the people to understand the dimensions of the crises other than just dwelling on what may be called "the superficial dimensions."

The story in Ahmed Yerima's award winning play, *Hard Ground,* revolves around the issues in the Niger Delta. The play is a contribution to the nation's sociocultural discourses and in specific context, the Niger-Delta crises. It is a simple account of the intense family relationships and the attendant emotional feelings sustained in precarious control by the threat of violence that surround the encounters by the various characters.

The actions in the play are easily identified by the reader. Nimi could be any son growing up and embroiled in the crises in the Niger Delta. Based on the appalling state of affairs in the society, he becomes hot-blooded, rebellious and above all, suffers from the conflict of identity with his father at the center and confused not knowing clearly the reason for the violent struggle he finds himself partaking in. Mama is the quintessential woman, sister and wife caught in the web of crises with her family torn apart in the process. Uncle Inyigifaa is like all those Chiefs in the land that have sold their integrities to the multinationals to the detriment of their communities for a morsel of food. The play explores deeply the link between the characters, events, settings and action to show that they all integrate to influence the sociopsychological aspects of life and hence will remain the same theme in any discussion on the issues in the Niger Delta.

In the play, characters are used in such a way as to portray a people who are envious of one another, always involved in intrigues and intricacies that work against each other. This way, the people never know who their real enemy is. The playwright also craftily handles the plot to successfully project the people as their worst enemy even as they look beyond their local area for the people who are working against their interests. By revolving the story around one family, *Hard Ground* ends up proving that one's real enemy is always one's best friend.

Further to the above, the play also examines pointedly the trend and pattern of agitation and militancy that now defines existence and politics in the Niger Delta. The play understands that violence or any form of raw action can achieve very little of the persuasive accuracy needed to resolve the issues that have necessitated the struggle in the first place. To highlight a fundamental slant of the havoc over the crisis for material resources is wrecking on human resources, the play positions the growth and development of the youth in the Niger Delta in social perspective. The average youth is portrayed in the play to be running a derailed psyche that renders them vulnerable to environmental pressures, which persuades them to either drop out of school or join any of the growing number of military outfits who claim they are fighting for justice. The play succeeds to cast its searchlight beyond the heavy rhetoric of oil politicking to represent the domestic, rural family and selfishly motivated actions prevalent in the society. But that the youth are potentially a lost group in the Niger Delta indicates a great loss not only in the oil war but also the corporate growth and development of the region as an entity. The Report of the Panel of Judges for the annual Liquified Natural Gas Prize for Literature supports this assertion when it states the focus of *Hard Ground* succinctly thus:

> The writer deflects the play's focus from the political and economic genesis of the Niger-Delta, to the emotional and spiritual impact on characters, and the shattering effect on close family ties. *Hard Ground* is a powerful and psychologically wrenching tale that draws on the sad

> history of the Niger-Delta as well as its contemporary, real-life political drama. It also gives us the occasional but telling glimpse into the pathos of personal relationships and the drawn-out psychological suffering of its most vulnerable characters. . . . But above all this, what the play effectively presents as the most valuable lesson of this war is the impact of the crisis on private lives.[23]

From the foregoing, it is noted that *Hard Ground* is not only an anatomy of the Niger-Delta imbroglio, a statement about the ongoing struggle and persistent issues surrounding the exploration, exploitation and distribution of the natural resources that have left the indigenes of the area in a state of neglect, squalor and abject poverty, but most importantly of all, it x-rays the treachery, corruption and greed at play in the struggle to possess their birthright. It shows that the pattern and trend of campaigning and militancy that now defines the existence and politics of the people of the Niger-Delta region go beyond socioeconomic and environmental theorizing by analysts. The play shows that all efforts to resolve the problems in the Niger Delta have failed because such efforts do not take cognizance of the controlling human factors that are sociopsychological in nature. Hence, the Niger Delta has virtually remained a cauldron of contradictions and a festering sore for Nigeria. By the above, Yerima has been able to interpret the dimensions of the crises in the Niger Delta through clothing it in its true human face and concludes by bringing out the real import, impact and the exigencies of the Niger-Delta crises that may lead Nigeria to perdition.

CONCLUSION

The chapter has shown using three plays written by Nigerians why analysts and the Nigerian government have displayed utter ignorance and paralysis at discussing the dimensions in the Niger-Delta crises. Therefore, they have found it expediently impossible to resolve the perennial factors that have conditioned the pervasiveness of the crises. The chapter did not seek to disprove all other discourses on the sources and dimensions of the problems in the Niger Delta but showed that the perspectives that advise on the dimensions thus far used in such analysis today need to be widened to include the sociopsychological issues and dimensions that are expressed in lucid and logical ways in recent literatures on the Niger Delta.

In light of the above, the chapter believes that the very root of the conflict in the Niger Delta is subsumed in the concept of security, which is the bottom line in every sociopsychological discourse. On one hand, local communities see the current pattern of the exploration, exploitation and distribution of oil with its devastating effects on the environment as a threat to the security of livelihoods. Government and those engaged in the oil business on the other hand see security in light of uninterrupted production and the marketing of oil, maybe, irrespective of adverse environmental and social impacts. The situation has been that neither the military nor the civilian governments have been able to forge a consensus necessary to manage these conflicts of securities. In spite of all assertions, the act of repressions employed by governments has simply made the Niger Delta ungovernable. Although oil companies and the government continuously made concerted efforts to resolve the issues through community and people orientation projects, these have failed because those efforts actually exclude the real people whose feelings and aspirations have always been wrongly gauged; hence, continuous violence and disaffection has continued unabated.

Policy makers may not be far from being wrong when they insist that the problem in the Niger Delta is really money. For example, the elite will continue to push for money and contracts. This has often been at the expense of the community. Upon obtaining these from the powers that be, it impresses on the rank and file to go for their share using whatever it takes to achieve this. However, beyond the interests of the elite, there are fundamental issues for the need to rewrite the history of the Niger Delta to reflect their emerging sense of nationalism and patriotism, participation and social rights, especially as it affects the God endowed natural resources, to be interpreted and resolved in a way to affect positively both the individuals and communities. Interestingly, from the onset, the problems of the Niger Delta cannot be interpreted simply by conjecturing that the problems exist because of the absence of social amenities, failure to pay compensations, gas flaring and a devastation of the hydro system in the Niger Delta. Similarly, looking at the crises from the dimension of the failure of the government to address the fundamental question of the federation, fiscal control, position of ethnic minorities, poor governance, end to militarism, implementation of the derivation principle/formula, an end to retrogressive laws, provision of job opportunities and so on and so forth will amount to discussing the nature of the dimensions superficially.

This chapter concludes that such discussions must transcend those above to include an understanding of the place of the people and their authentic representatives to share in real decision making with government and the explorers, exploiters and marketers of oil in addition to fathoming the sociopsychological dimensions of the feelings and behaviors of the people in the Niger Delta. This will show that if the many scares of many years of impoverishment, denigration, underdevelopment, inferiority complexes, greed, and all those factors that will help the people to assume their places in the comity of human beings are appropriately codified as postulated in the plays explored, it would have been realized that assessing the dimensions of the crises in the Niger Delta go beyond all others to include sociopsychological dimensions.

NOTES

1. John Pepper Clark, *The Wives Revolt* (Ibadan, Nigeria: University Press PLC, 1991).
2. John Pepper Clark, *All For Oil* (Ibadan, Nigeria: Malthouse Press Limited, 2000).
3. Ahmed Yerima, *Hard Ground* (Ibadan, Nigeria: Kraft Books Limited, 2006).
4. Jennifer Leaning and Sam Arie, “Human Security: A Framework for Assessment in conflict and transition,” USAID (December 2000), 4.
5. Paul Marris, “The Social Construction of Uncertainty,” in *Attachment Across the Lifecycle,* ed. C. M. Parkes, J. Stevenson-Hinde, and P. Marris (London: Routledge and Keagan Paul, 1991), 16.
6. Clement Ikpatt and Glenn Scott, “The Niger-Delta Problems and Solutions: The Equilateral Resource Control (ERC) Mode as an Alternative Dispute Resolution (ADR) Concept,” http/www.nigerdeltacongress.com/narticles/niger_delta_problem. Accessed 01/12/2009.
7. Okechukwu Ibeanu, “Oiling the friction: Environmental conflict management in the Niger-Delta, Nigeria,” in *Environmental Change and Security Project Report* 6 (Summer 2000): 21.
8. Oliver A. Opara-Nnadi, “Religion and conflict management in the Niger-Delta” (lecture delivered at the Theological Week of the Catholic Institute of Africa [CIWA], Port Harcourt, April 12th, 2006), 2.
9. Emeka Nkoro, “Conflict in the Niger-Delta: The Way Forward,” http/searchwarp.com/swa200447.htm. Accessed 01/12/2009.
10. Clark, *The Wives Revolt.*
11. Pat Utomi, “The Niger-Delta crisis: Beyond the price of crude oil,” http:/forums.crisis.org/Africa/?p=166. Accessed 01/12/2009.
12. Okechukwu Ibeanu, “Oiling the Friction,” 22–23.
13. Olu B. Osadolor, “The Niger-Delta Question: Background to Constitutional Reform,” Programme on Ethnic and Federal Studies (Ibadan, 2004), 9.
14. Report of the Panel of Judges for the Nigeria LNG Prize for Literature, http:/www.nlng.com/NLGnew/nigeriaprize/Report+Of+The+Pan. Accessed 04/22/2009).
15. Clark, *The Wives Revolt,* 59–60.
16. Ibid., 1.
17. Ibid., 58–60.
18. Clark, *All for oil,* 1.
19. Ibid., 15.
20. Ibid., 4.
21. Ibid., 22.
22. Nduka Nwosu, “Bekederemo Foundation and Museum: In Kiagbodo, Bekederemo Lives,” in http:/www.nigerdeltacongress.com/barticles/bekederemo_foundation. Accessed 04/22/2009.
23. Report of the Panel of Judges for the Nigeria LNG Prize for Literature, http:/www.nlng.com/NLGnew/nigeriaprize/Report+Of+The+Pan. Accessed 04/22/2009.

13 Creativity and Nature

Fagunwa's Tradition and Environmental Sustainability

Arinpe Adejumo

INTRODUCTION

Literary creation is influenced by the social milieu of the respective society. The development of society is a major concern for literary artists; therefore, literary works address their social and political climate. The Yoruba novel was born out of the attempt of the Christian Missionary Society to evangelize the Yoruba community and the efforts of Yoruba indigenes who desired to promote the Yoruba cultural heritage through literature.[1] The attempt germinated to different categories of novels. Ogunsina (1992) and Isola (1992) classified the Yoruba novel into four categories: the novel of Fagunwa's tradition, social novels, historical novels and crime novels.[2]

The novel of Fagunwa's tradition has some distinctive features that distinguish it from all other literary forms. It makes use of metaphors and symbolic animate and inanimate objects to convey ideas. More often, it uses satire to unveil evil being perpetrated in society, which the novelist feels can hinder growth and development. Among Yoruba novelists, D. O. Fagunwa is widely recognized for straddling the Yoruba marvelous and grotesque to bring about a magical realism. His success attracted some early novelists like J. O. Ogundele, Olu Owolabi, C. L. Adeoye and Akin Omoyajowo to write in his style. In all, there are about twelve novels in this category; the majority of them are written by Fagunwa, the pioneer of this tradition.

Eight of the novels shall be examined in this chapter to put the discussion of the millennium development goals in its proper perspective. They are Fagunwa's *Ogboju Ode Nínú Igbó Irunmole* (1950), *Igbó Olódùmarè* (1949), *Ìrìnkèrindò Nínú Igbó Elègbèje* (1954), *Ireke Onibudo* (1949), *Àdìtú Olódùmarè* (1961), L. O. Ògúndélé's *Ibú Olókun* (1956) and *Ejìgbèdè Lónà Isálú Òrun* (1956) and Olu Owolabi's *Oríí Adé Kìí Sùnta* (1974). The arrangements of the plots of these novels are similar. Yoruba prose narratives can be broadly divided into oral and written forms. The novel falls into the written form because it was imported from the Western literary tradition into Yoruba society before it was domesticated by novelists to suit the culture, beliefs and customs of their society.

One of the characteristics of folktales and stories in Yoruba is the use of archetypal characters to depersonalize the satirical portrayal of people and events to moralize and sanitize society. The folk tradition of oral tales is a weapon of reformation and ideological control. Also, through the use of folk tradition, socialization and internalization, society's values are reinforced. Most often, erring members of society are ridiculed in Yoruba folktales through the use of archetypal characters like tortoise, the trickster. Also, there are times that social commentaries are freely embedded in the composition of tales.

Novelists of Fágúnwà's tradition take cognizance of the storytelling techniques mentioned above in the composition of their literary works. The presentation of the novelists' point of view tows the line of Cooper's opinion that African novelists make use of animism to project their hopes and aspirations for a better society. Isola views presentation of events in the Fágúnwà's novels as a mixture of marvelous and grotesque to set a fantastic realism.[3] The characteristic features of the novels of Fágúnwà's tradition crystallize Cooper's view that

> African writers very often adhere to this animism, incorporate spirits, ancestors and talking animals in stories both adapted folktales and newly invented yarns, in order to express their passion, their aesthetics and their politics.[4]

The interspatial relationship between humans, non-humans and spirits is not alien to the Yoruba; therefore, the belief in supernatural forces, medicinal charms and magic occur in the novels of Fágúnwà's tradition freely to show the reality of life.[5] The presentation lends credence to the magical realists' view that magical realism is based on looking beyond the limit of the knowable to bring forth the truth of life.[6] Yet the truth of Fágúnwà's tradition novels could be captured only with 'the third eye.'[7] A rereading of novels in this tradition with the third eye reveals that novelists project far into the future of Nigerian society, and this is also applicable to the global world and the efforts they are making to ensure that the Millennium Development Goals are achieved.

PROJECTION OF THE MILLENNIUM DEVELOPMENT GOALS

The Millennium Development Goals (MDG), a pet project of the United Nations aims at meeting the needs of the world's poor, dates back to the year 2000. The project according to Kofi Annan, the former United Nation's Secretary, is achievable because it is people oriented and based on global partnering. He enjoins everybody concerned to play their roles in seeing to its achievement, as he asserts:

> I commend this report as a key resource . . . the analysis and information contained here can help citizen, organization, government, parliament and international bodies to play their respective roles in making the Millennium Development Goals a reality.[8]

The statement reveals that the achievement of the MDG is a joint responsibility of all.

Literary art is universally referred to as a representation of human experience. Therefore, advancement for the recent advocated MDG has been projected in the novels of Fágúnwà tradition. Symbolism is a main device employed to project the novelists' vision of the MDG. It is pertinent to note that the novelists in this category belong to an elite class in society. They have been exposed to the outside world either through reading or traveling to some developed countries; for instance, Fágúnwà had the opportunity of traveling to the United Kingdom.[9] These novelists were products of a hybrid culture. There was a great deal of cross-cultural comparison in their works, and this makes the projection of a global solution to the world's problems possible in their novels. Generally, at the surface level, the novels of Fágúnwà's tradition are adventure stories of heroes going on missions. In some of the novels, the hero is accompanied by other characters; for instance, Olówóaiyé in *Igbó Olódùmarè* is accompanied by other brave hunters in his exploit; also in *Ejìgbèdè Lónà Isálú*, Ọ̀run, Ogèèrè and Láńgbódókò accompany *Ejìgbèdè*; but in *Ibú Olókun*, Ọ̀rògodogànyìn goes alone on his journey. The journey embarked on by each protagonist is symbolic. It is an allegory of life's journey and the attendant problems associated with it.[10]

The settings of these novels are symbolic of the social, political and religious setting of the Yoruba people. The novelists express the intrinsic strand of modernity as it affects their enabling society. However, culture is not seen as being static, neither do they have an attitude of uncritical acceptance to all aspects of Yoruba culture. To their mind, culture is capable of change and growth. Their opinion has rightly been observed by Bamgbose while commenting on Fágúnwà's attitude on culture. He asserts that where customs that are contrary to growth and development are being portrayed, Fágúnwà deliberately ridicules such and highlights the standard values and culture.[11] He uses artistic license to launch his ideological belief. There are times that the novelist advocates for cultural diversity in the society. Although the Fágúnwà's tradition novels draw immensely from Yoruba culture, some of the visions projected go beyond Yoruba society. A deeper understanding of the novels reveals that the contemporary eight Millennium Development Goals have long been indirectly portrayed in the novels.

The Millennium Development Goals are the world leaders' declaration 'to address the issue of peace, security, development, human rights and fundamental freedom as a single package'[12] The eight MDG goals are to eradicate extreme poverty and hunger; achieve universal primary education; promote gender equality and empower women; reduce child mortality; improve maternal health; combat HIV/AIDS, malaria and other diseases; ensure environmental sustainability; and develop a global partnership for development.

The first MDG that is central to all the novels in Fágúnwà's tradition is the goal to encourage a global relationship for development. The UN advances a global social compact between developed countries and developing countries. Developing countries are encouraged to work out their development and these efforts are to be complemented by developed countries. This type of social compact is advanced in the interspatial relationship between the gnomes and fairies that the protagonists and their teams always experience in these novels. The essence of good international relationships is revealed in the way the protagonists embark on a journey that will lead to the development of their towns in the novels. To show that a global social compact is necessary, the protagonist will either go alone from home or meet other brave hunters in the bush or will go with other brave hunters from the outset. In Fágúnwà's novels, all the trips that the protagonists embark on alone are not always successful; but the end result is always successful when they go along with others.

The Yoruba people's belief in a global social compact is conceptualized in their belief in the relationship between the people in the world (*ayé*) and the ancestral spirits in the other world *(orun)*. Drewal also observes the cooperation between the world and other world as she asserts that

> In Yoruba thought, the other worldly domain (*orun*) co-exists with the phenomena world of people *(ayé)*. *Orun* includes a pantheon of uncountable deities *(òrìsà)*, the ancestors (*osi, egun*), and the spirits both helpful and harmful. The world and the otherworld are always in close proximity and both human and other spirits travel back and forth between the two.[13]

The above view encapsulates the fact that nations will achieve little development without cooperation and partnering. Thus, the issue of partnership pervades the novels under study. For instance, in Ògúndélé's *Ibú Olókun,* the protagonist goes to visit Àjàlórun, the Creator and owner of heaven, to acquire more knowledge. Also, in Fágúnwà's novels, the protagonists and their team always make a visit to beings living in heaven or some beings that are half human and half spirit, and as a result they have access to heaven. The role of the beings is to educate them on the path to progress and development for their towns. The combinations of the team that always go on the adventurous trip also support the view that partnering is crucial to the development of any group of people. Several instances could be drawn in the novels to support this claim. In *Ibú Olókun, Òrògodogànyìn,* the protagonist, embarks on his journey from home alone, but immediately he gets to heaven; *Àjàlórun* organizes characters like *Sùúrù (*Patience) and *Ìforítì* (Perseverance) who are to partner with him on his journey.

The essence of partnership for the development and growth of nations is also projected in Owolabi's *Orí Adé Kìí Sùunt,* Fágúnwà's *Ogbójú Ode Nínú Igbó Irúnmalè, Igbó Olódùmarè, Ìrìnkèrindò Nínú Igbó Elègbèje and Ìréké Oníbùdó.* In *Orí Adé Kìí Sùunta,* Òkòkòmeki Ajímutábàaoògun and two hundred others heroes join Kókóró ayé on a journey to Gbekuba, on the instruction of Obayem*i*, the king of Igbodola. In the same novel, Ògèrè, Láńgbondókò and Èjigbèdè go to heaven. Their combined efforts help them to surmount all the obstacles along the way. Also, in *Ìrìnkèrindò Nínú Igbó Elègbèje,* Ìrìnkèrindò (the adventurer) is called on by the king to lead some other great hunters and heroes to Òkè Ìrònú (hill of reflection) to go and pluck Igi Ìrònú (tree of reflection). Ìrìnkèrindò realizes he cannot accomplish the task alone. He then travels to the neighboring towns and cities without hesitation to find the great heroes. The heroes embark on the journey in unity of purpose because they are eager to liberate their society from underdevelopment, and they are ready to maintain peace. The efforts made by these great heroes could be likened to the world leaders' declaration to liberate their people from underdevelopment.

Furthermore, the heroes in *Ìrìnkèrindò Nínú Igbó Elègbèje* are able to solve many problems on their journey because of their cooperation and unity of purpose. They are able to deliver Ebora Ìgbéraga Ìbànúje (The spirit of pride) and Ahóndiwúrà (The tongue has turned to gold) from their predicament with cooperation.[14] Also, in the same novel, Ìrìnkèrindò and his team support the Èdìdàrè town, and the town is delivered from her crude way of life. This is a reflection of the way the developed nations could support developing countries as declared in the MDG. Eradication of extreme hunger and poverty is also one of the MDG goals. Reports show that "since 1990, millions more people are chronically hungry in sub–Saharan Africa and in Southern Asia."[15] Even in Nigeria, the Government is aspiring to reduce the poverty level to the barest minimum; recently, a national poverty eradication program was promoted to meet the citizens' needs. Hence, the way to combat poverty is a concern for all; thus, this topical issue is given attention in Yoruba literary works. Therefore, in these novels, solutions are proffered for hunger and extreme poverty.

The solutions offered by the novelists are based on a Yoruba maxim that says Ise laa se ka to jare osi (One needs to be hardworking to overcome poverty). The Yoruba also believe that knowledge and wisdom are essential in breakthroughs in life. Hence, Fágúnwà in his novels proposes that to be hardworking is not enough, but hard work must be complemented with wisdom and knowledge. In *Àdìtú Olódùmarè,* Fágúnwà recommends hard work, self-reliance and self-employment as one of the solutions to poverty. Also in *Ibú Olókun,* Ògúndélé supports the above by presenting a school of Renewal of Knowledge where an *Ole-ko-ri-kèngbè-da* (A-lazy-man-cannot-pluck-a-calabash) teaches the protagonist and other students the importance of hard work and knowledge. The teacher's name in Ibú Olókun is both symbolic and metaphoric. The process of plucking 'kèngbè,'[16] a type of calabash among the Yoruba is more tedious than plucking other types of calabashes. From this premise, it is discovered that a comparison is drawn between a lazy man who wants to pluck kèngbè and someone who wants to be liberated from poverty. As it is difficult for a lazy person to pluck kèngbè, so it is for him to get out of poverty. It could then be interpreted that for a nation to uproot the stem of poverty, extra effort from the Government and the citizens is required.

In translating this to the contemporary society, it could be inferred from the point of view of the novelists that dependence on government for white collar jobs may not bring the desired solution to poverty, as it is seen in the Nigerian and other African and Asian societies. A clue to the essence of self-reliance and self-employment is portrayed in the character of snakes in ilu awon ejò (town of snakes) in Igbó Olódùmarè. Snakes in this town rear toad so as to be independent.[17] Also, in Àdìtú Olódùmarè, Àdìtú who hails from a poverty-stricken family becomes a strikingly wealthy man, whereas his parents died of hunger.[18] Àdìtú in his adventure becomes animal friendly, and, thereby, he takes interest in nature research. At the end, the sale of elephant tusks makes him a wealthy man.[19] Àdìtú's action could be likened to the Obasanjo's National Poverty Eradication Program (NAPEP), a poverty alleviation program that encourages the youth to create jobs for themselves. Àdìtú in Àdìtú Olódùmarè later becomes a force to be reckoned with in his society as he puts in place many foundation and developmental projects.[20] Moreover, an assessment of Àdìtú's and the snakes' actions in the care of animals reveals that Fágúnwà proposes a state of environmental sustainability in society, which invariably may be a means of getting out of poverty as a nation.

The MDG goal of ensuring environmental sustainability aims at putting an end to deterioration of the natural resources and ozone layer, the availability of amenities like safe and drinkable water, bathrooms and toilets in every home.[21] As the United Nations is seriously agitating for a healthy environment for all, the novelists have long projected it in their arts.

All the novelists mentioned here have a vision for a well-structured society. The town of Èdìdàré could be an archetype of a typical unplanned and dirty environment. Èdìdàré town gives a picture of a town where goods are displayed in an open and dirty environment, houses are constructed without architectural plans, normal conveniences are not incorporated in the building designs and rooms are not well ventilated. The images from the description of this town fit into the construction of many houses in an average African society. Fágúnwà picturesquely portrays Èdìdàré town to deride the way town planning is generally handled in Nigerian society. Also, in Èjìgbèdè Lónà Ìsàlú Òrun, Ògúndélé describes Igbó Ìrójú as a place that is well maintained to the extent that it is more beautiful than any Oba's palace. The beauty of the place is described as follows:

> Igbó Ìrójú dára. Mo gbó wí pé Olódùmarè fí ewà se ibè ní òsò ní tewéko-tewéko, tènìa-tènìa, tigitigi igbó yìí ní o léwà . . . bí ènìà ba si wo igi ibè, kò ní wu olúwarè kí ó se e dáná nítorí pé ó dára púpò.[22]

> Igbo Iroju is beautiful. I heard that God beautifies the place; the plants, people, and trees in this forest are beautiful . . . if one looks at the tree in the forest, one would not want to cut them as firewood because of its beauty.

Even in the novel, the protagonist, Èjìgbèdè and his friends, Ògèrè and Lágbóndókò, are punished for killing an antelope at Òkè Olúgbódi.[23]

In addition, Fágúnwà also sees an unclean environment as a mark of foolishness and weakness on the part of the government. Thus, in *Ìrìnkèrindò Nínú Igbó Elégbèje,* the city of the people who are carefree about their environment is symbolically referred to as Ìlú àwon èdìdàré (the town of the fools). The symbolic name of their king is Òmùgòdiméta (foolishness multiplied threefold). The name of the king and his town are metaphoric. The government and people who subject themselves to living in an unplanned and unhygienic environment are sarcastically referred to as fools. The sarcasm is captured in the character of Òmùgòdiméta who is a weak and dirty king. The heroes on adventure are more enlightened than him; they persistently wage a war against him and his people until everything becomes normal. The war against dirtiness in the novel could be likened to the attempt being made by the United Nations, the Nigerian government and various agencies against the deterioration of the environment. For instance, there is great awareness of the need for a sustainable environment in contemporary Nigerian society. This brought about the inauguration of the State Waste Management Board in some states of the federation. Environmental sanitation laws were enacted to back up such programs. Also, for National orientation, the effort being made by the board is usually relayed on television stations, and documentaries are written for the state's public bulletins.

The above MDG goals dovetail into the MDG aims of reducing child mortality, improving maternal health and combating HIV/AIDS, malaria and other diseases. The MDG report findings reveals that 30,000 children die daily from preventable diseases, whereas more than half a million women die during pregnancy and childbirth.[24] To the average Nigerian such deaths are usually attributed to mystical forces and the belief that they are born in children. But the United Nation is making a frantic effort to put an end to the global premature and untimely death of children and pregnant women. Such efforts include the immunization of children and pregnant women along with campaigns and crusades against HIV/AIDS in every part of the world. Because the literary artists are part of the world, and literature is also people oriented, probable solutions to untimely death are addressed in the novels under study.

The effort of these novelists is a futurologist approach; at the same time, it could complement the government's effort. For instance, awareness of different kinds of diseases is created by the fairy named Ìránsé Olódùmarè (Creator's servant) in Ìrìnkèrindò Nínú Igbó Elégbèje. The fairy hawks all kinds of diseases in the novel; she educates the people on the causes of the diseases, and according to her, the people are the architects of their predicaments. She also showcases the reduction, prevention and cure of these diseases. She makes mention of Alátúnse (the rectifier) who buys the diseases from her. Alátúnse symbolizes a laboratory scientist who uses a dígí-pelebe (microscope) to diagnose the causative organisms of various kinds of diseases. The result of the tests will be given to the pharmacist who prepares medication for various diseases, and the medical doctors prescribe the medication for human beings.[25] The activities of the fairies could be compared to the contemporary activities of agencies like the World Health Organization (WHO) and the Global Alliance for Vaccine and Immunization.

The care of children is also one of the primary concerns of agencies like United Nations International children's fund (UNICEF) and WHO. Fágúnwà in his novels also airs his view on the care of women and children. According to Fagunwa the care of a child starts with the fetal stage. The novelist's opinion is revealed in the union between Olówóayé and Àjédìran in *Igbó Olódùmarè*. Olówóayé takes good care of Àjédìran when the latter becomes pregnant. He describes a person that fails to take care of a pregnant woman as a dead person and a viper:

> Eni tí kó bá se ìtójú aboyún kúrò ní ipò ènìà, òn ní à bá máa pè ní Paramólè tí ó gbé awo ekòló bóra: ejò olóró ni paramólè sùgbón Jéjé ni èkòló n sire tirè kiri.[26]
>
> He who does not take care of a pregnant woman is not a human being such a person would have been called a viper that puts on an earthworm skin: a viper is a dangerous snake but an earthworm crawls about harmlessly.

The issue of combating HIV/AIDS is not a preoccupation of the novels because the novels predated the HIV/AIDS scourge; all the same, the novelists warn against any act that can lead to sexually transmitted diseases. Research proves that indiscriminate sexual behavior promotes the spread of HIV/AIDS and other forms of sexually transmitted diseases. Hence, the slogan 'zip up' has been used in Nigeria to discourage indiscriminate sexual behavior among the youth. The slogan, although contemporaneous, has been implied in the attitude of the traditional Yoruba society: premarital sex is forbidden.[27]

Immoral acts are censored during various verbal performances at ritual and festival periods. Adulterers and adulteresses are ridiculed because an average number of citizens believe that anyone who gets involved in an adulterous act has violated the norms of the society.[28] The ridicule aims at discouraging the rate of sexually transmitted diseases and unwanted pregnancies. The novelists also corroborate the above belief and attitude toward adulterers, as they deride them in their novels. For instance, in Ibú Olókun, sexual chastity is promoted when Òrògodogànyìn (Pukish-fairy-in-mahogany-tree), the protagonist, proposes to Sùúrù (Patience) to be his wife. Sùúrù clearly reveals the Yoruba stand against adultery as she refuses the proposal instantly: "Ó ní òun kìí se omo àbí-ì-kó omoge abi yèrì pìrí, tí máa ta penpe lé okùnrin tì ó bá da enu ko ó láti fé"[29] (She says, she is not an untrained lady that quickly removes her underskirt, who jumps after proposals from men). The imagery of the act of removal of underwear by ladies during sexual acts is formed in the phrase 'abi yèrì pìrí' in the above excerpt. The opposite of 'removal of underwears' in this context is 'zipping up.'[30] It can therefore be implied that the novelists contributed their quota in the crusade against HIV/AIDS and other sexually transmitted diseases.

Human relationship at the gender level is a global phenomenon. Yoruba novelists also take the issue of gender and power relation as one of the themes to project in literary creation. Often, novelists take perception from the patriarchal angle. The issue of gender equality and women empowerment is also central to the MDG declaration. Feminism as a struggle in every part of the world is an organized movement for the liberation and emancipation of women.[31] Although, the struggle for women has taken a long time, the MDG report still shows that there is a long-standing inequality between men and women. In recent times, feminism developed into various groups like integral feminists, Christian feminists, radical feminists, liberal feminists and literary feminists, which are a few examples.

Looking at the novels of Fágúnwà's tradition, the portrait of women is based purely on the Yoruba cultural setting. In the Yoruba setting, women are portrayed in both negative and positive ways. There are no female protagonists in the novels. The gender politicking in the representation of women in the novel corroborates the United Nations' declaration that there are gaps between the representation of men and women at all levels.[32] All the same, it is discovered that female characters play key roles in the success of all the protagonists in the novels. Women in the novels are portrayed as helpers who come up at every point in time to save the protagonists and other heroes from danger. For instance, Olówóayé's dead mother comes to his aid in Igbó Olódùmarè. Also, Sùúrù and Ìforítì help Òrògodogànyìn in Ibú Olókun. The protagonists also get married to emphasize the binary complementary role between a man and a woman.

The equality of male and female genders is clearly revealed in the way Fágúnwà opposes the Yoruba custom that prevents a lady to propose to men. In his view, if a man can propose to a lady, nothing prevents a lady from expressing her love to a man. The above opinion is expressed in Igbó Olódùmarè when Àjédìran (Witchcraft-has-become-an-inheritance) proposes to Olówóaiyé by saying

> Iwo arẹ̀wà okùnrin, máse rò mí sí alábùkù nípa ìbéèrè tí ng ó se, mó fé kí ó rántí pé àsà burúkú àwon omo ènìà kò fún obìnrin ní àyè láti yan oko tí okan re fé, àwon okùnrin ní ó ní àyè àti ko wí fún obìnrin wí pé kí ó fé àwon. Bí okùnrin kǐan wú obìnrin láti fé àti so ó jáde lénù a di ìsòro fún obìnrin. Àsà àwon omo ènìà yí ńfí ìyà je apákan nínú àwon omo ènìà béni bí okùnrin rí ejò bí obìnrin pa ejò, bí ejò kò bá ti lo kó lòdì sí òfin.[33]
>
> Handsome man, do not take me for a despicable person because of the question I am about to ask. I want you to remember that the bad culture of human society does not give room for a lady to choose a husband for herself, it is only men that are allowed to propose to a lady, if a lady loves a man, she dare not say it. This culture is punishing a group in the human society, however, if a man sees a snake and the snake is eventually killed by a woman, the most important thing is for the snake to die.

A mockery of the tradition that prevents the female folks from exercising their right as regarding freedom of speech foregrounds the scenic presentation of the encounter between Olówóaiyé and Àjédìran in the above excerpt. From the above, it is found out that the issue of human rights as it pertains to the average woman is being emphasized by the novelist. The issue becomes a focal point in Fágúnwà's novels because in *Ìrìnkèrindò Nínú Igbó Elégbèje,* a princess, also proposes to a man she loves dearly.[34] Both proposals in the examples above end in marriages.

Ògúndélé in *Èjìgbèdè Lónà Ìsálú Òrun* equally focuses the complementary role of women and men and the emancipation of the female gender. The indispensability of male and female is conceptualized in the way Ogere, a woman, is able to overpower Olúgbódi after Lágbóndókò, a man, fails in this task. Ogere could not have achieved this had it been that she is not spiritually empowered.[35]

The significance of women's emancipation as a way out of women's subjugation is revealed in the relationship between Àdìtú and Iyùnadé in *Àdìtú Olódùmarè.* Àdìtú proposes to Iyùnadé's; he thinks she will readily accept the proposal owing to his wealth, but to his amazement, she turns him down. Iyùnadé acts like this because she is also economically empowered. Equally, a woman in *Ìrìnkèrindò Nínú Igbó Elégbèje* is portrayed as being economically empowered; this affords her the opportunity of bluffing the subjugation of men. Fágúnwà's view in the above examples reveals clearly that there is gender inequality in some aspects of Yoruba culture, and because culture is dynamic, the novelist proposes a change. In his view, women's right should not be infringed on; they should be given the freedom of expressing themselves. He also recommends that women are to be self-assertive and self-actualized if the society must truly experience growth and development. It could also be inferred from the portrayal of women in these novels that Fágúnwà, through his novels, also proposes a gender balance of empowerment for the benefit of the entire society because empowerment of women goes a long way in the achievement of other goals like the reduction of maternal death at childbirth and the reduction of child mortality as well as the acquisition of basic universal education for an average child. The advantage of women empowerment is rapped up in the MDG report that says

> Education, especially for girls, has social and economic benefits for society as a whole. Educated women have more economic opportunities and engage more fully in public life, As mothers, they tend to have fewer and healthier children who are more likely to attend school. All of these benefits are keys to breaking the cycle of poverty.[36]

The MDG aim of giving the average child the advantage of acquiring universal primary education indirectly crops up in the novels of Fágúnwà's tradition. In all the novels, the protagonists and other functional characters that accompany them usually go for informal education. For instance, the protagonists and their crew at the end of their journey usually end up going to a sage, who, through various stories, imparts into them the accepted norms of society. In Ògúndélé's *Ibú Olókun* there is a change in style of acquiring formal education. In the novel, there is a formal school at Ile-Eko Imodotun (The School of Renewal of Knowledge), which teaches that the basic norms of the people can enhance the growth of the nation. Vices like stealing and covetousness, which can lead to insecurity of the lives and property of the citizens in any society, are derided. All the novelists are cognizance of the fact that knowledge is power.

The essence of having a basic education and the role government and the populace can play in achieving this is reflected in Fágúnwà's *Àdìtú Olódùmarè*. Àdìtú provides scholarship for the brilliant students of Ilakose.[37] Also, in Ori Ade Kii Sunta, the fact that Kókóró Ayé (the key of earth) is educated enables him to become a very important personality. The educational empowerment also affords him the opportunity to write his autobiography, which forms the bulk of the novel. In the same novel, the skeleton of Alérànjà-Lógbún-òrun (one-who-pursues-goat-in-the-valley-of-heaven) is recognized easily at Iju Gbékúba because of the write-up attached to it by Alérànjà-Lógbún-òrun.[38] The education acquired does not give room for a break in communication among the characters as they journey in the wilderness.

The interrelationship of literature and sociology has been critically pursued in this chapter. The social milieu of the society being portrayed greatly affected the literary works under study. The futurologist function of literary work is also seen in the relevance of the novels of Fágúnwà's tradition to the contemporary 21st century Millennium Development Goal. The realistic nature of the novels lends credence to the fact that viewing the magical realist novels through the third eye could lead to the unveiling of the evil of our time.

NOTES

1. Bisi Ogunsina, *The Development of the Yoruba Novel* (Ilorin, Nigeria: Gospel Faith Mission, 1992), 6–15, 58.
2. Bisi Ogunsina, *The Development of Yoruba Novel,* 76, 100, 110; Akinwumi Isola, *The Modern Yoruba Novel: An Analysis of the Writers' Arts* (Ibadan, Nigeria: Heinemann Educational Books, Nigeria, Plc., 1998).
3. Akinwumi Isola, *The Modern Yoruba Novel,* 154.
4. B. Cooper, *Magical Realism in West African Fiction* (London: Routledge, 1998), 40.
5. Ayo Bamgbose, *The Novels of D.O. Fagunwa: A Commentary* (Ibadan, Nigeria: Nelson Publishers Limited, 2007), 7–18; Abiola Irele, "Tradition and the Yoruba Writer" *ODU* 11 (1975): 75–100.
6. A. B. Chanady as cited in Cooper, *Magical Realism,* 40.
7. Cooper, *Magical Realism,* 1.
8. Kofi Annan, *The Millennium Development Goals Report* (New York: United Nations, 2005), 3.
9. Ayo Bamgbose, *The Novels of D.O. Fagunwa,* 2.
10. Oso Olubusola, "Symbolism in D.O. Fagunwa and J.O. Ogundele's Novels" (master's project, University of Ibadan, 2006), 19.
11. Ayo Bamgbose, *The Novels of D.O. Fagunwa.*
12. Kofi Annan, *The Millennium Development Goals,* 3.
13. Margaret Drewal, *Yoruba Ritual: Performers, Plays, Agency* (Bloomington, Indiana: Indiana University Press, 1992), 26.
14. D. O. Fagunwa, *Irinkerindo Ninu Igbo Elegbeje* (London: Thomas Nelson and Sons, 1954), 59.
15. Kofi Annan, *The Millennium Development Goals,* 6.
16. Kengbe is a type of Calabash that hangs high among eekan (grass); it takes an extra labor to first cut the harmful grass before plucking the calabash.
17. D. O. Fagunwa, *Aditu Olodumare* (London: Thomas Nelson and Sons, 1961), 142.
18. Ibid., 42.
19. Ibid., 36–37.
20. Ibid., 44–45.
21. Kofi Annan, *The Millennium Development Goals,* 30.
22. J. O. Ogundele, *Ejigbede Lona Isalu Orun* (London: Longman, 1956), 68.
23. Ibid., 64–65.
24. Kofi Annan, *The Millennium Development Goals,* 18–24.
25. D. O. Fagunwa, *Irinkerindo,* 26–27.
26. D. O. Fagunwa, *Igbo Olodumare* (London: Thomas Nelson and Sons, 1949), 54.
27. Nathaniel Fadipe, *Sociology of the Yoruba* (Ibadan, Nigeria: Ibadan University Press, 1970), 309.
28. Arinpe Adejumo, "Conflict Resolution in Oral Literature: A Review of Yoruba Satirical Songs," *The African Journal of New Poetry* 5 (2008), 108–109.
29. J. O. Ogundele, *Ibu Olokun* (London: University Press, 1956), 52.
30. 'Zip up' is a slogan used for mobilization and crusade against HIV and connotes total abstinence from premarital sex.

31. Aduke Adebayo, *The Nature and Function of Literature: The Comparatist's Perspective* (An Inaugural Lecture, University of Ibadan, 2009), 18–20.
32. Kofi Annan, *The Millennium Development Goals,* 14.
33. D. O. Fagunwa, *Igbo,* 26.
34. D. O. Fagunwa, *Irinkerindo,* 94.
35. J. O. Ogundele, *Ejigbede,* 66–67.
36. Kofi Annan, *The Millennium Development Goals,* 10.
37. D. O. Fagunwa, *Aditu,* 44.
38. Olu Owolabi, *Ori Ade kii Sunta* (Ibadan, Nigeria: Macmillan Nig. Publishers Ltd., 1974), 81.

Conclusion
Nigeria and the Global Environment

Toyin Falola and Adam Paddock

The recent oil crisis in the Gulf Coast offers pause to reflect on the relationship between government, entrepreneurial activity and the environment. Nations and businesses engage in a dialectic process that determines the rules of extractive industries and the limitations placed on them in terms of environmental dangers to local communities. Whereas the crisis in the Gulf Coast is an extreme example of the destructive potential of offshore oil drilling, there are some interesting comparisons between how governments decide to protect the environment and the possible effects of local crises on the global community. Just as important as understanding the causes of these environmental disasters is how they are dealt with years later in terms of cleanup and prevention.

Nigeria and the United States have legislation in place that is supposed to limit the chance for environmental hazards as well as hold responsible those parties who create environmental disasters. Yet the balance between free enterprise and government regulation is a fundamental issue that each nation faces. Conservative politics tend to favor little government regulation on businesses; however, more liberal politics argue for a measure of government regulation to prevent the indirect consequences of business activity. The decision between no government regulation and government intervention is more than an ideological or political stance but reflects specific ideas about the relationship between government and national economies. If nothing else, the recent crisis in the gulf shows how local national practices can have far-ranging affects.

Arguments in favor of limited or no government regulation believe that such regulation hinders the market economy by placing constraints on business activity. In economic terms, regulation alters the market supply curve by raising the production costs of goods. The increase in production costs is then reflected in an upward movement of prices, which leads to a movement along the demand curve and a drop in the demand of those goods at the higher price. The U.S. economy is largely dependent on mass consumption, and the contraction of market activity owing to a decrease in the amount demanded at the higher cost is often believed to slow production and economic growth. Milton Friedman was the chief proponent of free market capitalism in the 1970s, ideas that greatly influenced U.S. political economics for four decades. It is only with the most recent financial crisis that many U.S. economists are reevaluating John Maynard Keynes concept of government intervention during economic downturns via deficit spending to initiate economic activity.

Liberal politics in terms of environmental regulation argue that traditional market economic theory does not capture the complete costs of production. Supply and demand curves capture the demand consumers have for a product at specific prices and the willingness of businesses to supply goods at those prices. The supply curve is calculated according to the total costs of production including fixed costs and variable costs. Fixed costs include costs that remain relatively constant such as building rental fees. Variable costs include things such as supplies and wages. If the total costs of production rise higher than the current market price, the price must either rise or the business must end its ventures. The problem with classical theoretical supply and demand curves is that they do not exist in reality. In the real world many factors affect or manipulate free market enterprise and the relationship between supply and demand.

Fixed and variable costs captured in the supply curve do not account for the indirect costs of production. This is why liberal economists support environmental legislation to balance out the indirect effects of production that are not captured in free market capitalism. A classic example is the mining industry. Strip mines pollute water sources and cause erosion in areas outside the land occupied by the mining company. Pollution to the water table does not affect the production activities of company, but it can have dire effects on local inhabitants who become ill owing to water pollution. In this scenario either the local inhabitants suffer the consequences and must pay for their own medical treatment or the government must pay out of tax revenues. Regulation places rules on how mining companies operate so that they do not pollute local water supplies, or if they do, it places stiff penalties so they bear the full costs of their enterprise. Regulations and fines also protect citizens from paying the costs of cleanup and environmental degradation in the form of taxes by passing those costs onto the companies responsible.

The difference between the United States and Nigeria lies in a difference in the level of transparency, fragmented national politics and the system of doing business with international companies. The U.S. government by design enforces regulatory policies to prevent against environmental disasters, and when they do occur often a unilateral public response demands quick action. On the other hand, the fragmented nature of Nigerian politics and a high level of government kickbacks from the oil industry lead to an uneven enforcement of environmental legislation. The uneven enforcement or complete lack of any enforcement of environmental regulations in Nigeria means that local inhabitants are paying the indirect costs of production in their quality of life. Corruption in the form of company support for local politicians insulates legislators from the democratic process. Large sums of money from companies allow politicians to remain in office or new ones are won over to the companies' interests after election. To make matters worse, the fragmented structure of national politics means that legislators on a national level find no reason to support citizens other than those from their districts. With companies blocking local legislators' actions against pollution via kickbacks, local inhabitants find little recourse to address their situation.

Tropical environments are some of the most ecologically diverse regions on earth and countries in the developing world have to negotiate decisions between economic development and conservation. More often than not, and despite the outcries of local citizens, governments tend to favor short-term monetary gain over long-term questions of environmental sustainability. This was one of the main themes in the volume. The role of the Nigerian environment in the history of the pre-colonial, colonial or post-colonial periods reflects some similar issues; but there is also change throughout the region's history in the economic, social and political strategies that informed the relationship between the environment and the people. The legacy of the colonial experience is exemplified by broad cultural changes and affected the relationship between the environment and the people, which is one such change.

The pre-colonial relationship between Nigerians and their environment was not a utopian experience of perfect symbiotic balance. Nigerians made choices about how to alter their environment by choosing what agricultural goods to grow, how to smelt or mine minerals, how to grow crops, when to try growing new kinds of food crops and also where and how to create modes of transportation by using rivers and overland trade routes. In fact, as Akinwumi and Filaba showed, Nigerians possessed an intricate knowledge of their environment and they knew how to utilize herbs and poisons to heal and to kill, a method of self-defense and knowledge to which colonizers did not have access. Pre-colonial Nigerians may not have used the same type of scientific analysis that was popular in Europe but they possessed their own methods of analysis that led to important herbal remedies and medicines no less effective than European medical science at the time, which often led to bleeding as a remedy for illness.

The colonial period was a divisive period that without question led to profound changes in Nigerian culture, society, politics and economics, but also the environment as well. Yet the colonial influence should not be overstated because Nigerians made choices about how they wanted to negotiate the colonial experience and they did influence colonial policies. In many ways, the colonial period was defined by a contest between different sets of goals about how environmental resources should be used and who should benefit from those resources. The British strategy was to export as many extractive and agricultural goods as possible for sale on the international market. Nigerians, in many respects, were willing to benefit from the trade;

however, they were not passive but tried to determine the terms of sale. The palm oil trade mentioned in the introduction was one case of Nigerians protesting for better prices. Nigerian farmers who switched between cash crops and food crops also provide an example of how Nigerians attempted to maximize their earnings depending on market prices.

Nigerian cities existed in the Hausa region to the north and in the Yoruba region to the west before the onset of British colonial rule. Cities had also been established along the coast owing to the slave trade in the eighteenth century; however, British colonial rule did offer an impetus for urban migration owing to a new set of opportunities that many rural inhabitants found more appealing than farming. With the new opportunities of growing cities came new problems of infrastructure. The relationship between the people and the environment was changing, and owing to an inability of the British government to keep up with city improvements to match the growing population, environmental problems arose that are still with Nigeria today. Undependable electrification, trash and sewage problems all affect life in the cities in one way or another.

One of the most debated questions in Nigerian history is to what extent contemporary problems in Nigeria are owing to the colonial legacy or ineffective post-colonial governance. In terms of the environment, this question remains pertinent; and as the previous chapters showed, it is not easy to answer. British technology had an important impact on the environment in a variety of areas including mining techniques, farming and transportation, each of which affected the environment to varying degrees. Yet post-colonial Nigerian governments largely failed to effectively correct the environmental and infrastructure problems left by British colonialism, and to some extent those problems have been exacerbated in the last twenty years.

The oil industry figured prominently in the volume and for pertinent reasons considering that it accounts for a large percentage of Nigeria's GDP and because it is the main environmental issue in the country, which has led to extensive violence, political unrest and deprivation in the Niger Delta. Yet Kairn Klieman's chapter illustrated how divisions of time that historians often use are general and that there is not a sharp divide between colonial and post-colonial Nigerian history. It was a period of transition in economics and politics, but the affect independence had on Nigerian's environmental policies appeared to be minimal. Many of the same colonial practices continued and international markets in part favored Nigeria's economy to remain predominantly extractive in nature. The global free market influence of economically powerful countries like the United States and China appear to overshadow the ability of local inhabitants to effectively influence their political leaders to respond to their needs. Economic globalization is about more than a transfer of ideas and goods from one region of the world to another, but it represents an era where local political power and representative government can be greatly influenced by external markets and political forces. This situation presents grave dangers to the democratic process in developing countries and in the case of Nigeria is one of the primary causes of combatants' actions in the Niger Delta.

Oil has been Nigeria's blessing and curse. It has provided the country with much needed income, but at the same time how the money has been used is a topic of debate that has led to virulent criticisms. The deprivation of people living in the Niger Delta is well documented in the previous chapters; the affluence of those benefiting from the oil industry is in stark relief to the suffering in the Delta. The environmental degradation caused by the oil industry is largely a post-colonial problem, but the issue is also related to the colonial experience. The general question is one of power. Oil multinational companies are empowered by the Nigerian government and operate with few impediments, even though the actions of the companies have been irresponsible and at some times criminal. Niger Delta discontent is related not only to the fact that they do not benefit from oil revenue and that their environment is being destroyed but also because their government fails to protect them from abuse by foreign companies.

The nexus between environment, economics and politics most often deals with broad issues of governance and international markets and how global influences affect local inhabitants. Yet that is only half of the issue. The Nigerian environment and the challenges it faces are also about how people see themselves in terms of their environment, how that has changed overtime and why it is important. Environmental issues and Nigeria's extractive focused economy is not limited to GDP and international commerce; it is also about how people's relationship with their environment changes. Environment is part of a people's identity; people identify with specific landmarks, which have cultural and religious significance. The connection between land and identity is a strong one. It was common for Nigerians of many ethnicities to bury family members under their homes. This was part of the process of remembrance and identity creation; yet, it is environmental degradation that for many low-income citizens in Nigeria challenges their ability to maintain connections of identity, either in urban slums or the Delta region.

The preceding chapters illustrated a variety of ways in which Nigerians negotiated the restraints of the environment, colonial rule and post-colonial governance issues to manage their access to natural resources. In the pre-colonial period, Nigerians farmed a variety of foods including new agricultural goods from South America. They also developed techniques for utilizing natural resources such as iron, gold mining and other minerals. During the colonial period, they used the colonial system to their advantage when possible and in other instances circumvented or directly challenged colonial authority. Charles Ukeje's chapter evaluated how the people in the Niger Delta dealt with those very issues during the colonial and post-colonial periods. Several chapters based on Nigerian literature also illustrate the hopes and anxieties that people have concerning their environment and the problems of development in the late twentieth and early twenty-first century.

The goals of Nigerians are no different than the rest of Africa or the world in general: they seek to carve out the best lives that they can for their families. Whereas environmental conservation is a concern throughout the world, many Western countries like the United States do not deal with the same high level of poverty and underdevelopment. To add to the problem, as several chapters have discussed, how oil revenue is used presents at least a crisis in governance and perhaps a human rights issue. Historians debate the extent of colonial influence since independence and whether or not international markets' operations are neo-colonialist. Just as pertinent to the discussion, in terms of the Niger Delta, is whether it is possible that the affects of sporadic military dictatorships in Nigeria have left the remnants of a flawed system. Many Niger Delta activists seem to think so.

The environmental problems in Nigeria, as bleak as they seem at times, have not been without heroes. As tragic as the death of Ken Saro-Wiwa was, he brought international attention to the Delta and spurred investigations into the activities of oil companies and the complicity of the government. Unfortunately, the environmental tragedy is not over for Nigeria or many other African nations. The government and people continue to face difficult challenges ahead as they combat poverty and growing environmental degradation that affects the quality of living for many people. Nigerians, like the citizens of all countries, affect and shape the environment in which they live; but at the same time, we are all victims of the environmental problems we create.

Contributors

Jamaine Abidogun is currently an Associate Professor in History at Missouri State University at Springfield. She holds a PhD from the University of Kansas in Curriculum and Instruction with a minor in African and African-American Studies. She has published several chapters and articles based on ethnographic research in Ghana and Nigeria. She was awarded a Fulbright Scholarship in 2004–2005, which resulted in her article, "Western Education's Impact on Northern Igbo Gender Roles in Nsukka, Nigeria." Her research focuses on education's role in cultural change in the United States and West Africa. She is a current member of African Studies Association (ASA) and the Mid-America Alliance for African Studies (MAAAS). Dr. Abidogun has taught African history courses at Missouri State University and at the University of Nigeria–Nsukka.

Arinpe Adejumo is a poet, playwright and literary critic, and a Senior Lecturer in the Department of Linguistics and African Languages, University of Ibadan. She obtained a PhD in Yoruba Language and Literature from Obafemi Awolowo University, Ile-Ife. Her academic background and research interests include studies in satirical genre, gender studies and folklore studies. She has over 25 publications to her credit. Her publications articulate the interdisciplinary interface between literature and other cognate disciplines vis-a-vis the function of literature in society's growth and development.

Olayemi Akinwumi is a lecturer with the Department of History, Nasarawa State University, Keffi, Nigeria. He received his PhD in African History from the University of Ilorin, Nigeria. He has taught courses in African History in Nigeria and abroad. He is a recipient of many international awards. Some of these are the prestigious German Alexander von Humboldt award, the German Educational Exchange Award (DAAD), etc. He has published in many international journals. His recent book, *Conflicts and Crises in Nigeria: A Political History,* is published by the LIT Verlag, Germany. His research interests are intergroup relations and African indigenous science and knowledge systems.

Dele Ashiru is a Doctoral Candidate in the Department of Political Science, University of Lagos, Nigeria. He teaches The History of Political Thought, Political Economy, Political Behavior, Methodology of Comparative Politics and the Politics of Development and Underdevelopment in the same University. His recently published articles include *"The Judiciary and the Democratization Process in Nigeria"* and *"Chieftaincy Institution and Grassroots Development in Nigeria."* His current research interest is in the area of political Islam, fundamentalist pressures in Africa and the political economy of development.

T. T. Asojo is a retired Associate Professor and former head, Department of Religious Studies, Faculty of Arts, Olabisi Onabanjo University, Ago-Iwoye, Ogun State, Nigeria. She holds an MA in Philosophy of Religion and a PhD in Philosophical Foundations from the University of Ibadan, Nigeria. She was the first African woman award winner of the Science and Religion Course program of the Center for Theology and Natural Sciences (CTNS), Berkeley, California, in 2001. This award facilitated the introduction of a BA degree in Religion and Science course in Olabisi Onabanjo University and was also one of the seven pioneer winners of the Olabisi Onabanjo University Distinguished Service Award in 2004. She was visiting Professor in the Women Studies program of the University of Oklahoma, Norman, in 2004 and 2007.

Her research areas include philosophy of religion, science and religion and women's studies.

She has authored two books and over thirty journal articles and has also presented numerous papers on science and religion, philosophy of religion and women's studies at international conferences in the United States, United Kingdom and Nigeria.

Mailafiya Aruwa Filaba is Professor at Nasarawa State University, Keffi, Nigeria. He teaches history of Nigeria and Latin America. He specializes in urbanization, human rights, environmental issues and cultural history. He has authored and coauthored several books.

S. U. Fwatshak obtained his PhD in the History of Entrepreneurship in Jos, Central Nigeria, from the University of Jos in 2003, and he is a history lecturer at the Department of History and International Studies, University of Jos, Nigeria. During the 2000–2001 academic session he was Fulbright Fellow at New York University, New York City. His research interests span contemporary African economic and social history in which he has carried out research and published several articles. His most recent publications include *African Entrepreneurship in Jos, Central Nigeria, 1902–1985* (Durham: Carolina Academic Press, 2010). Between 2007 and 2009, he was Nigeria Country Coordinator of the Volkswagen Foundation, Germany-sponsored *Sharia Debates Project* following his active involvement as a researcher on the Volkswagen Sharia Debates Project (2003–2004), and Cordaid, Netherlands-sponsored Sharia Implementation in Nigeria Project (2005–2006).

S. O. Ijaola NCE (Agric. Education DM), BTh (Hons), PGDE, MBA (Agribusiness), MA (Philosophy of Religion Major) is a PhD candidate of Christian Religious Studies specializing in a Philosophy of Religion Major with a focus on Religion and Science in the Department of Religious Studies, Olabisi Onabanjo University, Ago-Iwoye, Ogun State, Nigeria; and a Faraday Institute Course (Religion and Science) alumnus at St. Edmunds College, Cambridge, United Kingdom. He has been involved in teaching of Religion and Science and Philosophy of Religion in Olabisi Onabanjo University since 2007 to date. He was both a curriculum consultant and Lecturer at Pentecostal Sanctuary Theological Seminary. His robust academic interests revolve round the fields of agriculture, education, theology, poverty, business administration, philosophy of religion, environment, religion and science, sociology and African traditional religion. He is presently carrying out academic research in philosophy of religion, religion and science and environment and has his current coauthored and authored publications in the *Association of Study of Interplay between Religion and Science (ASIRS)* and *Ayigba Journal of Arts and humanities (AJAH)*. He among others is an ordained Reverend of Nigerian Baptist Convention and had served in various churches; He is also a management consultant, motivational speaker and investment analyst.

Kairn Klieman is an Associate Professor of History at the University of Houston. She received her training in pre-colonial African history at UCLA, and her first book, *"The Pygmies Were Our Compass": Bantu and Batwa in the History of West Central Africa, Early Times to c. 1900*, (2003) was a finalist for the Herskovitz Award and a CHOICE Outstanding Academic Title. She is currently at work on a book entitled *Before the "Curse": Petroleum, Politics, and Transnational Oil Companies in the Gulf of Guinea, Africa, 1890–1980*, which provides a detailed history of relations between transnational oil companies, their home governments and the colonial and independent governments of three "Gulf of Guinea" States (Democratic Republic of Congo, Nigeria, and Angola). Dr. Klieman has received numerous awards and fellowships in the course of her career. For earlier research she received support from the Social Science Research Council, the Fulbright-Hayes Foundation, and the Belgian-American Educational Foundation. More recently she received the University Teaching Excellence Award, one of the highest honors bestowed by the University of Houston.

Alexander Kure is a PhD who lectures at the Department of English and Drama at Kaduna State University, Kaduna, Nigeria where he is also the Director, General Studies. His major research interests are in applied linguistics and comparative literature as it interrogates gender, peace/conflict resolution and environmental issues. He is a member of the Linguistic Association of Nigeria (LAN), Nigerian English Studies Association (NESA), and English Language Teachers Association of Nigeria (ELTAN).

Samuel Ojo Oloruntoba is a PhD Candidate in the Department of Political Science, University of Lagos. He holds a Masters of Science Degree (with Distinction) in Political Science from University of Lagos, Nigeria, with specialization in Politics of International Economic Relations. He is also a Chartered Secretary of the Institute of Chartered Secretaries and Administrators, (ICSA), United Kingdom. He has both industry and academic experience spanning over a decade. He has held several Associate Consultancy positions with international development organizations like the International Trade Centre, an UNCTAD/WTO ITC organization in Geneva, Switzerland and the World Bank. In 2009, he was a Visiting Fellow at the Watson Institute of International Studies' (Brown International Advanced Research Institute on Development Studies), Brown University, in Rhode Island. He is a Member of the Network of Excellence on Qualitative Research Methods in Africa (QRMNet) an EU-EDUlink Program anchored by the Institute of Development Studies, University of Sussex, United Kingdom. He teaches International Relations, Political Economy, Politics of Development and Underdevelopment and African Politics at the University of Lagos, Nigeria. Oloruntoba has written and presented papers at conferences in his areas of Research Interest such as Democratic Governance, the Global Trading System, Human Security, Migration, Global Capitalism, State Capacity, Civil Society and International Organizations. His latest published work by the *Nigerian Journal of Business and Social Sciences* is "The Global Economic Crisis and the Challenges of Human Security in Africa." He appears regularly on television programs and grants newspaper interviews on issues of political, economic and social relevance to Nigeria's development. Samuel Oloruntoba is currently a Visiting Scholar at the Program of African Studies, Northwestern University, Evanston, Illinois.

Donald O. Omagu received his PhD from the University of Calabar, Calabar, Nigeria, and currently teaches Global and African History at Wagner College, New York. He is the author of *A History of Bekwarra People of the Upper Cross River Region Of Nigeria* (1997) and *Regional Peace and Security: A Historical Perspective of the Role of Ecowas in Liberia and Sierra Leone* (2001). His scholarly writings have appeared as chapters in edited volumes. His research interests include Nigeria and West Africa.

Michael Sharp (PhD, University of Wisconsin–Madison) is professor of English & Caribbean Studies at the University of Puerto Rico. He has taught in Scotland, Greece, Portugal, Nigeria and at Binghamton, Harvard and Wisconsin; he has also been a Faculty Associate in the School of Theory in the Humanities at the University of Santiago de Compostela, Spain. Professor Sharp's research centers on the poetry of English-speaking Africa and the Anglophone Caribbean.

Charles Ukeje is a Reader in International Relations, Obafemi Awolowo University, Ile-Ife, Nigeria, and until recently held a Lectureship in African Politics and Development in the Department of International Development, Queen Elizabeth House, University of Oxford, England. A past recipient of the prestigious Fulbright Scholarship, he was given the Mary Kingsley Zochonis Lecture Award of the Royal African Society, United Kingdom, in September 2004, and the Leventis Cooperation Visiting Research Scholarship at the Centre for African Studies, SOAS, London, the following year. He has published extensively in major peer-reviewed journals, including *Issue-A Journal of Opinion*; *Africa Development*; *Politics, Culture and Society*; *Journal of Terrorism and Political Violence*; *Oxford Development Studies*; *Ethnic and Racial Studies*; and *Globalisations*. He coedited *The Crisis of the State and Regionalism in West Africa: Identity, Citizenship and Conflict* (2005) and contributed a chapter in Colette Daiute et. al., *Global Perspective on Youth Conflict and Resilience* (New York: Oxford University Press, 2006). He also coauthored *Post-War Regime and State Reconstruction in Liberia and Sierra Leone*, Comparative Research Network Report (Dakar: CODESRIA Book Series, 2009), as well as a chapter titled "Changing the Paradigm of Pacification: Oil and Militarization in Nigeria's Delta Region," in Cyril Obi and Siri Aas Rustad (editors), *Oil and Insurgency in the Niger Delta: Managing the Complex Politics of Petroviolence* (London: Zed Books, 2010). Dr. Ukeje is currently one of the key facilitators of the Executive Postgraduate in Peace and Security in Africa, a Program of the African Union in conjunction African Peace and Security Program of the IPSS, University of Addis Ababa, Ethiopia.

Index

For Product Safety Concerns and Information please contact our EU
representative GPSR@taylorandfrancis.com
Taylor & Francis Verlag GmbH, Kaufingerstraße 24, 80331 München, Germany

www.ingramcontent.com/pod-product-compliance
Ingram Content Group UK Ltd.
Pitfield, Milton Keynes, MK11 3LW, UK
UKHW022116190926
475702UK00013B/1478
9780415847421